OCEAN POLLUTION

Effects on Living Resources and Humans

The CRC Marine Science Series provides publications that synthesize recent advances in Marine Science. Marine Science is at an exciting new threshold where new developments are providing fresh perspectives on how the biology of the ocean is integrated with its chemistry and physics.

CRC MARINE SCIENCE SERIES

Series Editors

Michael J. Kennish, Ph.D.
Peter L. Lutz, Ph.D.

Published Titles

Ecology of Estuaries: Anthropogenic Effects, Michael J. Kennish
Ecology of Marine Invertebrate Larvae, Garry McEdward
Morphodynamics of Inner Continental Shelves, L. Donelson Wright
Physical Oceanographic Processes of the Great Barrier Reef, Eric Wolanski
The Physiology of Fishes, David H. Evans
Practical Handbook of Marine Science, 2nd Edition, Michael J. Kennish

---

Forthcoming Titles

Benthic Microbial Ecology, Paul F. Kemp
The Biology of Sea Turtles, Peter L. Lutz and John A. Musick
Chemical Oceanography, 2nd Edition, Frank J. Millero
Chemosynthetic Communities, James M. Brooks and Chuck Fisher
Coastal Ecosystem Processes: Temperate vs. Tropical, Daniel M. Alongi
Environmental Oceanography, 2nd Edition, Tom Beer
Major Marine Ecological Disturbances, Ernest H. Williams, Jr. and Lucy Bunkley-Williams
Marine Bivalves and Ecosystem Processes, Richard F. Dame
Practical Handbook of Estuarine and Marine Pollution, Michael J. Kennish
Seabed Instability, M. Shamim Rahman
Sediment Studies of River Mouths, Tidal Flats, and Coastal Lagoons, Doeke Eisma
The Physiology of Fishes, 2nd Edition, David H. Evans

OCEAN POLLUTION

Effects on Living Resources and Humans

Carl J. Sindermann, Ph.D.
National Marine Fisheries Service, NOAA
Cooperative Oxford Laboratory
Oxford, Maryland
and
Rosenstiel School of Marine and Atmospheric Science
University of Miami
Miami, Florida

CRC Press
Boca Raton New York London Tokyo

Library of Congress Cataloging-in-Publication Data

Sindermann, Carl J.
Ocean pollution: effects on living resources and humans / Carl J. Sindermann.
p. cm. -- (Marine science series)
Includes bibliographical references and index.
ISBN 0–8493–8421–4
1. Marine fauna — Effect of water pollution on. 2. Seafood — Contamination. I. Title. II. Series.
QL121.S62 1995
591.52′636--dc20

95–16809
CIP

Direct all inquiries to CRC Press, Inc., 2000 Corporate Blvd., N.W., Boca Raton, Florida 33431.

International Standard Book Number 0–8493–8421–4
Library of Congress Card Number 95–16809
Printed in the United States of America 1 2 3 4 5 6 7 8 9 0
Printed on acid-free paper

Preface

Coastal/estuarine pollution has become recognized in recent decades as a significant and expanding problem. Events during the past several years have emphasized the need to understand the effects of man's activities on habitats and on the use of fish and shellfish resources. Information has accumulated, but is fragmentary, and it has not been evaluated in terms of total effects on fish stocks.

Human impacts on estuarine and coastal ecosystems increasingly seem to occur in direct proportion to the numbers of people in adjacent land areas, so that a map that plots human population distribution and density effectively locates major coastal and estuarine pollution problems as well, with the single caveat that the degree of industrialization of the populated area must also be considered. Just as an example, the waters adjacent to the extended New York metropolitan area are, and could be expected to be, among the most severely degraded in the entire world. Pressures from an adjacent human population of some 18 million, exerted in the form of ocean dumping, ocean outfalls, non-point source runoff from land, petroleum spills and leakages, and fallout of airborne contaminants — all act in concert to make the New York Bight and its associated estuarine areas a case history of degrading activities applied to the coastal environment.

Detailed long-term studies in this and other badly degraded coastal/estuarine areas such as the Southern California Bight, Puget Sound, the Thames estuary, the Baltic sea, Tokyo Harbor, and the Mediterranean Sea have provided a better understanding of pollutant effects. It is clear from these studies that major changes in coastal ecosystems have been caused by man's effluents. The full *extent* of these changes is still somewhat difficult to assess. Benthic populations have been modified, and many have even been eliminated in severely contaminated areas in practically every major harbor and developed estuary. Dissolved oxygen values near the bottom have been found in some polluted areas, at certain times, to be too low to support the normal life of bottom-dwelling fish. Toxic trace metals and petroleum contamination of bottom sediments have been measured at extremely high levels. A number of abnormalities in fish and shellfish seem related to the extent of degradation.

More difficult questions, such as specific effects of pollutants on fish and shellfish *abundance,* still remain unanswered, but sufficient information is available about the effects of coastal and estuarine pollution on living resources to warrant a reasonably dispassionate review of the subject. This volume is the result of that review, concentrating on data from the Atlantic coast of the United States, but including published scientific information from many other estuarine and coastal areas of the world as well.

The book has been in various stages of composition or decomposition for almost two decades (with occasional digressions to write chapters for other people's books, and even to write a separate book on the closely related topic of fish and shellfish diseases). It began modestly enough as an attempt to assemble what was known about pollution-associated diseases of marine animals, and then it moved outward to encompass quantitative effects of pollution on resource populations and public health aspects of coastal pollution. Somewhere in the writing process, considerations of scenarios for the future and strategies to reduce pollution levels were viewed as necessary components. Well toward the end of a first draft disillusionment set in over the intensely technical tone of the existing text — disillusionment strong enough to force a complete rewriting in an attempt to "lighten up" the approach and to make the material a little more accessible and understandable (and maybe of greater interest) to general readers.

In preparing this material, I have profited greatly from discussions with many colleagues in the United States and elsewhere. The deliberations of the Environmental Quality Committee of the International Council for the Exploration of the Sea and its related working groups have been particularly helpful in providing insights about coastal pollution in the North Atlantic and adjacent waters. Several recent workshops and symposia on coastal/estuarine pollution and its effects have also been sources of significant new information. Additionally, several edited volumes about the effects of pollution on coastal/estuarine environments have been published during the past decade, but none has examined the problem of coastal pollution from the viewpoint of cumulative effects on resource species, which is the perspective of this volume.

I would like to acknowledge in particular the great benefits of long-term discussions about coastal pollution with my good friend Dr. John B. Pearce of the National Marine Fisheries Service, Woods Hole, Massachusetts. Dr. Pearce came to the Sandy Hook (New Jersey) Laboratory in the early 1960s to study pollution effects on resources of the New York Bight. His many papers on pollution effects on benthic animals, his development of large-scale monitoring programs, and his emergence as an international authority in ocean pollution matters took place principally during the decades of the 1970s and 1980s. I was fortunate enough to be associated with him at that laboratory from 1970 to 1984 and to profit from the extended professional relationship with an exceptional marine scientist.

I would like to express my appreciation also to Mrs. B. Jane Keller, Editorial Assistant, Cooperative Oxford Laboratory, for her meticulous attention to the almost endless drafts and reorganizations of the book manuscript. Her expertise has contributed significantly to whatever literary merit the book may have.

Thanks are due to colleagues in the National Marine Fisheries Service and at the University of Miami, especially Dr. Aaron Rosenfield of the Cooperative

Oxford Laboratory and Dr. Edwin Iversen and Mrs. Electa Pace of the University of Miami, for many useful comments and suggestions (without, of course, implying their responsibility for any statements or omissions in the final document).

I would also like to acknowledge the hospitality of the Commonwealth of Massachusetts, for providing facilities for writing and contemplation at South Pond in the Savoy Mountain State Forest high in the northern Berkshires. Without drawing too many gratuitous parallels, South Pond is in many of its characteristics the present-day equivalent of the well-known, but now despoiled, Walden Pond (located in the eastern part of the Commonwealth) as it was more than a century ago during Henry Thoreau's tenancy there.

The book is written for a general audience, even though the concepts are not simple, some of the terminology may be a little obscure, and some of the arguments may seem at times just too technical. It tries to create a quiet zone in an era of environmental emotionalism, not by glossing over the available evidence, but rather by trying to evaluate that evidence calmly and without too much bias. To achieve this end, I would like to acknowledge the insights gained from a reexamination of the writings of the late Rachel Carson, especially her earlier books on the sea. Ms. Carson succeeded in combining lyrical, almost poetic, narrative with technically precise, factual content — her works are models for all of us who aspire to write in this genre.

Even a cursory perusal of the various sections of the book will disclose two that seem most important (at least as indicated by the almost infallible criterion of the number of pages devoted to each). These are (1) the associations of diseases — of resource species as well as humans — and levels of pollution, and (2) the relationship of pollution and resource population abundance. The reasons for preeminence of these two topics seem perfectly obvious, at least to me. The presence and severity of diseases are indicators of stresses on individuals in the populations at risk, and abundance changes can be reflections of the intensities of the stresses, whether they be natural or human-induced.

The structure created for the narrative, then, has these two major pillars supporting a flow of related but subsidiary information. I can see six definite phases to the book: (1) a consideration of pollution as a stressor and an overview of survival responses of marine animals to it; (2) an examination of diseases in fish and shellfish (and indirectly in humans) as consequences of pollution; (3) a review of some data about the effects of pollution on populations of fish and shellfish; (4) a speculative look at large-scale and small-scale marine events that may have some pollution linkages; (5) a brief description of some possible future scenarios in marine pollution; and (6) a few proposals for reducing coastal pollution. The extent to which cohesiveness is achieved within this framework for the presentation must of course be assessed by the reader.

Since few people ever read a technical book like this one from cover to cover (and rightly so, since it is, after all, not a novel), I offer several options:

1. Skip lightly through the italicized vignettes in each chapter and ignore the rest of the text. This approach will give a soupçon — a tiny taste — of the flavor and content of the entire document.
2. Be selective. Read chapters 2, 3, 5, and 7, then decide whether any more information can be tolerated.
3. Go for it! Read every chapter in sequence, word by word, looking up references at the ends of chapters as necessary adjuncts to the text material.

If option 3 is chosen, please be patient with a degree of deliberate redundancy. Some topics discussed early in the book beg to be revisited from another perspective in a later chapter, and I have occasionally listened to their pleas.

Carl J. Sindermann
Oxford, Maryland

The Author

Carl J. Sindermann, A.M.,Ph.D.,D.Sc.(Hon.), is currently a senior research scientist associated with the Oxford (Maryland) Laboratory of the National Marine Fisheries Service. Formerly director of marine research laboratories in Florida, New Jersey, and Maryland, he is the author of more than 150 technical papers in marine biology. He has written and published five books on the role of disease in marine populations (one of which received an outstanding publication award from the Wildlife Society of America), and has edited and published eight other technical books. In addition, he has contributed chapters to several books on food and drugs from sea and on marine aquaculture, and he served as co-chairman of a published symposium volume on diseases of fish and shellfish.

Dr. Sindermann has also written and published four non-technical books about science and scientists — *Winning the Games Scientists Play* (Plenum, 1982), *The Joy of Science* (Plenum, 1985), *Survival Strategies for New Scientists* (Plenum, 1987), and (with C. Yentsch) *The Woman Scientist* (Plenum, 1992). The first was cited by *Library Journal* as one of the best scientific/technical books of 1982. He is at present an adjunct professor at the Rosenstiel School of Marine and Atmospheric Science, University of Miami. He has held adjunct professorships at a number of other universities, including the University of Rhode Island, the University of Guelph, Cornell University, and Georgetown University; and was previously on the faculty of Brandeis University. He received his advanced degrees from Harvard University.

Dr. Sindermann has been active in international science. Until recently he was Chairman of the International Council for the Exploration of the Sea's Working Group on Introductions and Transfers of Marine Organisms, as well as a member of the United States-Japan Joint Natural Resources Panel on Aquaculture. He has served as scientific advisor to several international fisheries commissions and has participated in a number of international fisheries symposia and conferences. He earlier served as the Scientific Editor for the National Marine Fisheries Service and Editor of the journal *Fisheries Bulletin*. He is on the editorial boards of several scientific journals and has been a member of various National Sea Grant Site Review Panels. Earlier he was a member of the organizing committee of the Society for Invertebrate Pathology, and he served as a president of the World Aquaculture Society.

The Author

Dedication

This book is dedicated to my wife, Joan, who has always been my strongest supporter, especially during episodic writing sprees of the kind that has led to completion of this document.

Dedication

Contents

Illustrations

Vignettes

* Journal entry date

Prologue

THE NOT-SO-FANCIFUL TALE OF A HUDSON RIVER FISHERMAN

Once upon a time, not so long ago, in a village by an arm of the sea, there lived a fisherman and his family. The fisherman, and his father before him, and his father before him, had always looked to the coastal waters for their livelihood, catching fish and shellfish in season. The work was hard but satisfying; owning a boat and fishing gear provided a precious sense of independence despite a variable and at times only marginal income.

As the years passed, more and more outsiders decided that they would come to live in the fisherman's village. Housing developments and then condominiums spread across the fields, and new marinas lined the shores, polluted the harbor, and crowded the working fleet into moorings adjacent to the marshy, less desirable land. Commuters disappeared from the town into the distant city each morning, and straggled back in the evening; then spent the weekends in their power boats or tending lawns and gardens. The locals and the commuters achieved an uneasy balance of political power in the town, but the interests of developers were almost always served.

Then one day in early spring, 1976, just as the azaleas bloomed in his front yard, the fisherman's world collapsed. New regulations. Some of the species of fish that he depended on for his livelihood were declared "unfit for human consumption" by an anonymous bureaucrat in the state public health agency, and the taking of those fish was prohibited. According to reports in the New York Times, two manufacturing plants far up the Hudson — units of the General Electric Company — had been dumping toxic chemicals — especially polychlorinated biphenyls [PCBs] — into the river for decades, and enough of the obnoxious stuff had been carried downstream to contaminate the fish that inhabited the lower reaches. So-called "action levels" of the pollutants were found in the flesh of the fish, and the "action" was to close the fishery, as a protection for consumers.

The fisherman, and others like him, spent a long time alternating between frustration and rage — a repetition of their reactions during the previous summer, when the clam beds outside the town were closed to fishing because of increasing levels of fecal contamination from the expanding human population. He and his dispossessed cohorts felt increasingly mistreated and rejected by a changing society that gave no indication of respect for their way of life. He also felt control of his existence slipping away from him, almost capriciously and certainly without his consent.

But the response of that changing society to the fisherman's questions would be that there is no answer, except that he is an anachronism — a living relic of an earlier, gentler time when the human species had not yet achieved sufficient numbers to despoil the land and then foul the coastal waters. So-called "environmental awareness" appeared too late to save him from a land-based service job, or maybe one on some industrial assembly line — far from the pristine edges of the sea that exist only in his memory.

We do not need to identify the fisherman, but he does exist, as do many like him. The habitat degradation described in his story also exists, in a variety of forms and degrees of severity in different parts of the nation's coastline. This book is one regretful bow to a way of life that is disappearing — crowded out by demands of an expanding human population for "alternate uses" of coastal/estuarine waters. I hope, as undoubtedly he does, that it is really not too late — that gradually improving environmental value systems in America will halt and even reverse the long-term destruction of coastal habitats and their occupants, even in such people-overwhelmed areas as the Hudson River estuary, Boston Harbor, and Puget Sound.

From "Field Notes of a Pollution Watcher"
(C. J. Sindermann, 1978)

Introduction

Humans have existed as the species *Homo sapiens* for several hundred thousand years. For almost all of this period they could have been dismissed readily as "somewhat gregarious, mildly destructive, and generally slovenly terrestrial animals." Only during this century (a tiny fraction of one percent of their total tenure) have they begun to make any measurable impact on the oceans that cover most of the planet: since the Industrial Revolution as instigators of chemical and physical changes that have modified rivers, estuaries, and localized segments of coastal waters; and during the past four decades as serious predators on ocean fish stocks. This book is an attempt to summarize our present understanding of how contamination resulting from human activities — collectively termed *pollution** — has affected and is affecting living marine resources, especially those species that inhabit the vulnerable coastal/estuarine zones.

Reasonable evidence exists to demonstrate that pollution of coastal waters of the world is increasing. It does so as the human population and industries in coastal zones expand, as ocean outfalls proliferate in numbers and capacities, and as ocean dumping continues and increases in volume. Indicators of the problem include high coliform levels in coastal waters; increased organic content in inshore sediments and reports of coastal algal blooms; and reports of significant levels of chlorinated hydrocarbons, petroleum residues, and heavy metals in ocean waters, sediments, and organisms. The U.S. coastline, for example, is characterized by the presence of some badly degraded but localized waters that get much of the public's attention (the Hudson River estuary and Boston Harbor are good examples) and by much *relatively* clean coastal water. Generally, as might be expected, the degree of coastal/estuarine pollution is directly proportional to human population density and the extent of industrialization in the adjacent land areas. A similar conclusion has been

* It might be well at the outset to define the term *pollution*. It can be described succinctly yet broadly as "anything animate or inanimate that by its excess reduces the quality of living,"[1] or more narrowly and less succinctly as the "introduction by man, directly or indirectly, of substances or energy into the marine environment (including estuaries) resulting in such deleterious effects as harm to living resources, hazards to human health, hindrance to marine activities including fishing, and impairment of the quality of sea water,"[2] or, as an alternative, as "human activities causing negative effects on marine life, human health, resources, or amenities."[3]

reached by investigators in other parts of the world. An international working group on pollution in the North Sea reported that "the North Sea as a whole was not seriously polluted, but that areas of the coastline adjacent to high population and industrial centers had apparent problems that required further attention."[4] Predictions of population increases in the coastal zones of the United States and other countries suggest that pollution problems will intensify and spread.

Information about the effects of pollutants on living marine resources and on the ecosystems that support their production is also increasing rapidly. In examining this information, however, it is sometimes difficult to separate conclusions based on factual information from those based on inference or speculation. Currently, there seems to be extensive extrapolation of limited data by some media writers, who may then arrive at conclusions not always fully supported by the available facts. Such speculations and extrapolations are cited and often augmented by other environmental writers, creating a spiral effect, with the entire spiral balanced precariously on a very narrow data base. In marine pollution matters, as in other matters with high emotional content, we would be well-advised to approach conclusions conservatively and to not venture too far beyond the safe confines of available data. However, this does not imply that problems do not exist or that there is not some hard information available about the effects of ocean pollution on living resources; it is with this kind of information that this book is concerned. In general, the correlations between fisheries resource problems and coastal pollution may be categorized at different levels of credibility: (a) *substantiated,* with good documentation; (b) *suspected,* with weak documentation; and (c) *speculated or inferred,* with little or no documentation. Where clear associations have been established, this will be indicated; where only poorly documented or suspected relationships exist, this will also be indicated.

The degree of risk to living resources (and their utilization) from pollution represents a continuum from none to severe; these are some possible forms it can take:

1. Individuals may suffer physiological impairment, resulting in slow growth rates, lowered fecundity, and accelerated mortality.
2. Individuals and localized populations may be destroyed by pollutants.
3. Entire species may be reduced in abundance or even eliminated from parts of their ranges by acute or long-term impacts of pollutants.
4. Seafood may become a public health hazard because of the presence of microbial or chemical contaminants.
5. Consumer acceptance of seafood may be reduced because of off-flavors or fear of poisoning or disease.
6. Other economic losses may result from closure of shellfish beds because of microbial pollution, or fisheries in certain areas may be prohibited because of high contaminant levels in the product.

We can be frustrated in our efforts to understand pollution effects, because we lack good baseline data on previous abundance levels of resource and food chain organisms and on previous environmental chemical levels. In a few places — the New York Bight, Puget Sound, and the Southern California Bight, for example — we are at least acquiring *current* baseline data against which future changes can be measured; but historically we still must rely on scattered and incomplete records to assess man's impact on marine populations and ecosystems. The absence of good baseline information often relegates conclusions about environmental and population changes to the circumstantial rather than the proven category.

Although there may be a number of ways to compartmentalize the available information about marine pollution, for clarity and order in a book like this, I have elected to consider marine pollution effects on living resources from six perspectives:

Factors leading to survival or death in degraded habitats
Ocean pollution and disease
Effects of ocean pollution on populations
Perceptions of ocean pollution
The consequences of inaction
Proposals for change

I would emphasize at the outset that marine pollution problems are largely *coastal* and not *oceanic,* although some high-seas pollution exists. Affected areas include estuaries, coastal areas adjacent to estuaries, other coastal areas adjacent to municipalities or large industrial complexes, and (to a much lesser extent) the continental shelf areas further removed from the immediate coast. During the past two decades some attempts have been made and some successes achieved in reducing pollution loads in rivers and canals, but often these benefits have been achieved at the expense of bays, estuaries, and coastal waters; through increased marine sludge dumping; and the proliferation, consolidation, and extension of ocean outfalls.

I would emphasize also that the technology for the solution of coastal/estuarine pollution problems is available, but awaits suitable application of money and determination. Coastal waters have multiple uses (recreation, fishing, aquaculture, transportation, waste disposal, energy generation, and chemical extraction), with priorities for each in a state of dynamic and uneasy equilibrium. As an example, until recently, use of shelf waters as dumping areas had high priority. With the belated realization that man's wastes could seriously degrade coastal waters and living resources, this priority is being reexamined, but to date the costs of alternative solutions to ocean dumping have been overriding factors in the reassessment.

One observation that emerges from an examination of coastal pollution and fisheries is that while reasonable evidence exists for *localized* effects, there is as yet little specific evidence of *widespread* damage to major fisheries resource populations resulting from coastal pollution. There is also some evidence that other factors, such as repeated year-class successes or failures, long-term shifts in geographic distribution, and overfishing, may cause major changes in fish abundance.

Field investigations of possible relationships between pollution and localized damage to fish and shellfish populations usually yield evidence that may at best appear circumstantial. However, the weight of such evidence, when combined with results of experimental contaminant studies and scrutiny of available information about long-term trends, favors the conclusion that a relationship does actually exist. What is needed in the immediate future is continued and increased effort to test and document the reality and extent of the relationship. This book describes where we stand right now.

One major objective is to provide a damage assessment of the inshore marine environment and its inhabitants at a time when heightened environmental awareness characterizes public attitudes. Principal findings discussed herein are that acute problems exist in areas of the worst degradation, but these are localized, so examination of most coastal areas discloses less severe pollution damage to living resources than might be anticipated.

This is not to suggest that coastal habitat degradation has not occurred or is not occurring — or that we understand enough about population impacts of sublethal levels of pollutants to assess effects with much validity. In the few instances where data bases are adequate to permit realistic simulations of the effects of pollutants on population size, results have been varied. A few suggest identifiable population impacts; most others do not. The continuing reality seems to be that effects of specific pollutants on abundance of resource species have not been demonstrated adequately, in large part because of the difficulty in separating pollution effects from effects of overfishing, or from effects of variability in natural environmental factors. A quantitative area desperate for more data is that of the effects of pollution on growth and reproduction of fish and shellfish — effects that are extremely difficult to document, but are critical to understanding the total interaction between pollutants and the survival of resource populations.

Other perspectives on pollution damage, in addition to direct effects on resource population size, are explored in this book. These include disease effects (on humans as well as fish) and effects on aquaculture. Good reasons for concern lurk in each of these categories, and issues of public health as well as resource health are involved.

REFERENCES

1. **Spilhaus, A.** (Cited in Bascom, W. 1974. The disposal of waste in the ocean. *Sci. Am.* 231(2): 16–25.)
2. **National Academy of Sciences**. 1971. Marine environmental quality: Suggested research programs for understanding man's effect on the ocean. Report of a special study under auspices of the Ocean Science Committee of NAS-NRC Ocean Affairs Board, 9–13 August 1971. 107 p.
3. **Kinne, O. and Aurich. H.** (Eds.). 1968. Biological and hydrographical problems of water pollution in the North Sea and adjacent waters. *Helgol. wiss. Meeresunters.* 17: 1–530.
4. **ICES** (International Council for the Exploration of the Sea). 1975. Report of thc 7th session of the Group of Experts on the Scientific Aspects of Marine Pollution (GESAMP). ICES Doc. C.M.1975/E:11, 25 pp.

REFERENCES

SECTION I:
Pathways of Pollution Effects: Some Basic Concepts

To begin a discussion of pollution effects on living resources with any hope of success we have to grapple immediately with two interwoven basic concepts:

1. Pollution contributes to stress in marine animals—as an added environmental factor with which they must cope; and, of equal importance,
2. Coastal/estuarine animals are equipped with physiological/biochemical mechanisms that enable adaptation and survival in degraded habitats, unless the environmental changes are too drastic or too prolonged—in which case disability and death occur.

If in this section we can achieve some appreciation for the functioning of these two concepts in fish and shellfish at risk from pollution, the rest of the agenda for this book will make good sense. It seems logical to begin in Chapter 1 with an exploration of the nature and expression of stress in fish and shellfish, particularly as it may be induced by contaminant chemicals of human origin. Once this task has been accomplished, we can move on in Chapter 2 to a discussion of physiological/biochemical mechanisms promoting survival and well-being, or, if these mechanisms are overwhelmed, leading to disability and death in polluted habitats.

1 The Role of Stress in Marine Animals

****

STRESSED EELS FROM THE ELBE RIVER

The estuary of the Elbe River, near the German city of Hamburg, epitomizes many of the evils that an industrial society can inflict on the aquatic environment. The shores are lined for miles with shipyards, chemical factories, and wharves for oceangoing commercial vessels. The water is usually a dingy gray-brown, with abundant floating debris and high levels of chemical contaminants. Despite this overwhelming evidence of man's destructive presence, a marginal fishery for eels, flounders, and other species persists, but the prevalences of abnormalities — indicators of environmental stress — in those fish are high.

It was in this venue, with eels from these degraded waters, that a university researcher, Dr. Gabrielle Peters, reported, in 1980, results of a classic study of how environmental factors, including pollution, can produce stress in fish. She selected eels (Anguilla anguilla) for experimental work probably because they were readily available, as they were in many other European coastal waters. When placed in stressful environments, the eels exhibited many of the same physiological and morphological signs seen in stressed humans: the stomach lining thinned and ulcerated, blood glucose levels increased, and steroid hormone levels dropped. When kept under conditions of prolonged stress, some eels became emaciated and died within a few weeks.

Dr. Peters' experimental findings of stress-induced damage were augmented by field observations made by others. Eels captured in this polluted estuary (like those from other degraded European estuaries) exhibited numerous bacterial infections, precancerous changes in liver tissues, and massive cell proliferations in the head region, aptly called "cauliflower disease" (Figure 1) — all conditions thought to be associated with pollution.

Since eels are widely distributed in temperate zone countries, where they are usually considered gourmet-class seafood, and since they are now being cultured in Europe as well as in the Far East, these findings of stress-related defects have great relevance for the future. Young Atlantic eels travel great distances from their birthplace in the Sargasso Sea to reach European and North American shores, and after extended growth periods in fresh water the females again move into coastal waters to rejoin the males before their return spawning migration to the deep waters of the Atlantic. Abnormal environmental conditions such as gross chemical pollution in any of these successive habitats in continental waters can reduce the likelihood of successful completion of a truly amazing life history, that has to this day defied full scientific description.

From "Field Notes of a Pollution Watcher"
(C. J. Sindermann, 1995)

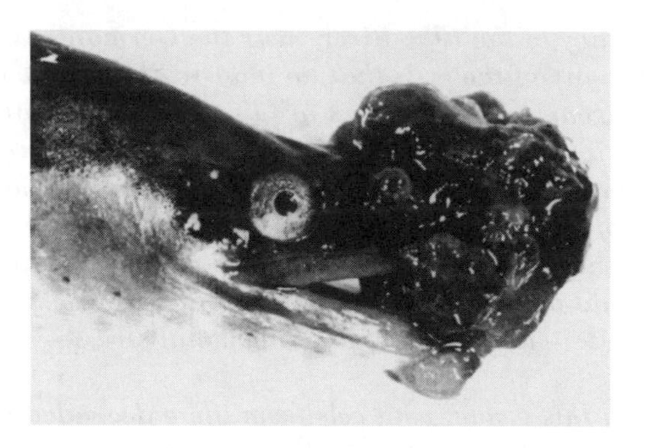

FIGURE 1. "Cauliflower disease" of the European eel.

****

Environmental factors that affect fish and shellfish can be identified, and their principal elements are listed in Figure 2. The diagram includes most of the principal sources of stress for aquatic organisms and also illustrates the extent of the problem before us. Some factors are biological, others are physical or chemical; single factors may dominate at any particular time. Some are directly lethal, others can lead to debilitation, physiological malfunction, or morphological abnormalities that render individuals more vulnerable to the effects of other factors.

It is obvious that pollutants and other man-made changes constitute only one of the total array of factors — physical, chemical, and biological — that impinge on fish populations. Pollutants and other nonoptimum environmental factors act as *stressors,* which, if extreme enough or prolonged enough, may

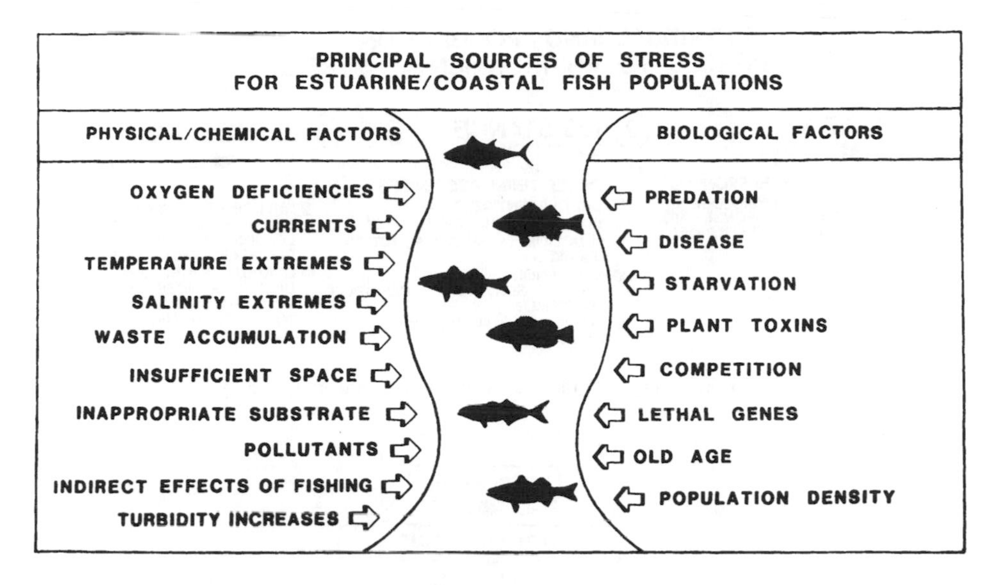

FIGURE 2. Environmental variables that affect fish stocks.

affect survival. *Stress* is a significant but elusive concept in biology, and, like any concept, it can be misunderstood, misinterpreted, or abused. In its original sense, as developed by the Canadian scientist Hans Selye,[1] *stress represents the consequences of all the mechanisms whereby an organism attempts to maintain equilibrium in the face of environmental change.** Selye's original description of stress emphasized physiological responses of the organism in three phases — *alarm, resistance,* and *exhaustion* (Figure 3). The alarm phase includes immediate or short-term behavioral, biochemical, or physiological responses to nonoptimum changes in the environment; the resistance or adaptation phase includes longer-term biochemical/physiological responses that improve the likelihood of survival in the nonoptimal environment; and the exhaustion phase includes failure of critical biochemical functions, leading to physiological and morphological disorders and death. The biochemical pathways and effects of stress are summarized (probably too briefly) in Figure 4.

* Many definitions of stress have been proposed. Selye defined stress as the sum of all the physiological responses by which an animal tries to maintain or reestablish a normal metabolism in the face of a physical or chemical force. Others have proposed many modifications of this definition — such as "a state produced by any environmental or other factor which extends the adaptive responses of an animal beyond the normal range, or which disturbs the normal functioning to such an extent that, in either case, the chances of survival are significantly reduced".[2] Another definition that clearly identifies stress as the product and not the cause of homeostatic change is, "Stress is the effect of any force which tends to extend any homeostatic or stabilizing process beyond its normal limit, at any level of biological organization".[3] All of the proposed definitions and descriptions of stress are vaguely unsatisfying, but in its broader sense *stress represents the sum of morphological, physiological, biochemical, and behavioral changes in individuals which result from actions of stressors.*

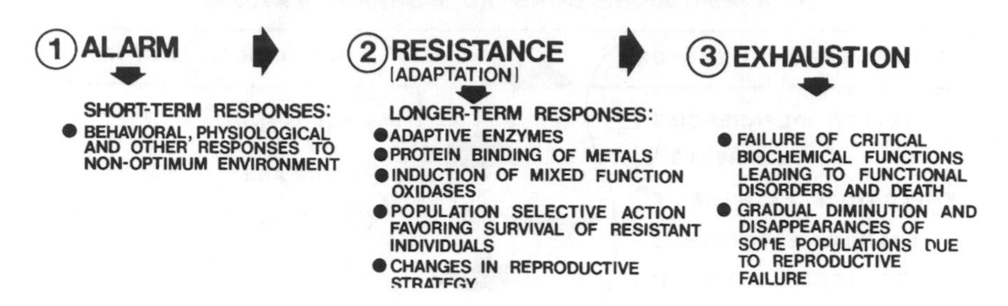

FIGURE 3. Stages in the classical stress syndrome of Selye.

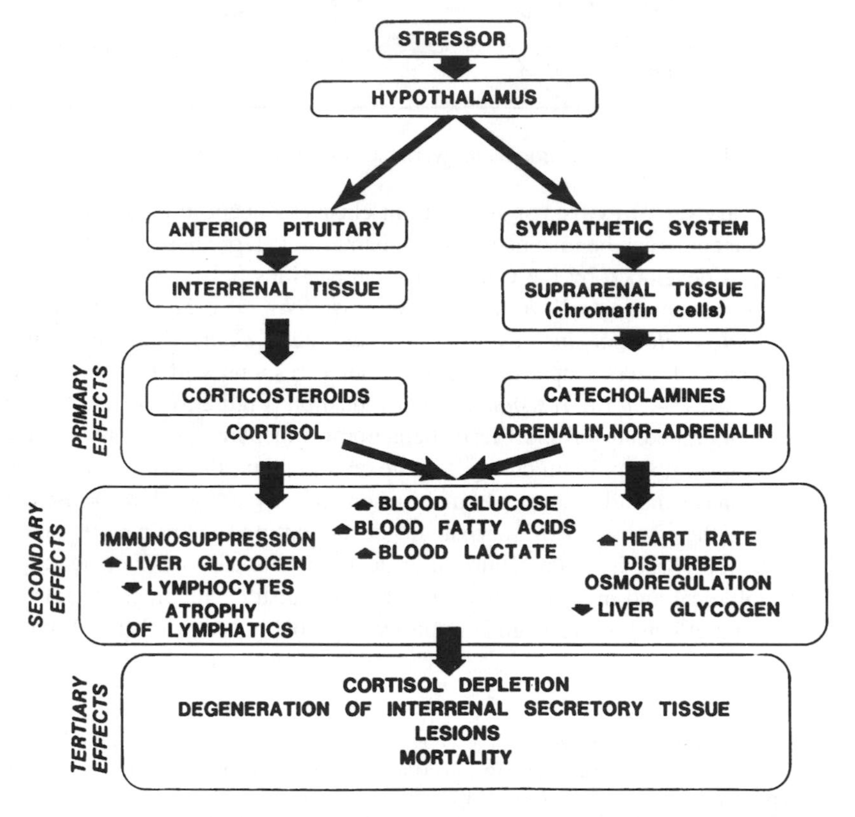

FIGURE 4. Pathways of the stress syndrome. (Modified from Mazeaud, M.M., F. Mazeaud, and E.M. Donaldson,[4] *Trans. Am. Fish Soc.* 106: 201, 1977.)

As shown in this somewhat complex but very important diagram, physiological/biochemical responses (accompanied of course by behavioral and morphological changes) include:

Primary: Increased output of corticosteroids and catecholamines
Secondary: A multitude of metabolic and osmoregulatory disturbances — among the most important of which are immunosuppression and decreased lymphocyte production
Tertiary: Decreased resistance to disease, increasing physiological malfunctions or abnormalities: effects on growth and reproduction, and death

These responses are distributed temporally; some occur immediately, whereas others may take months to develop, as illustrated in Figure 5.

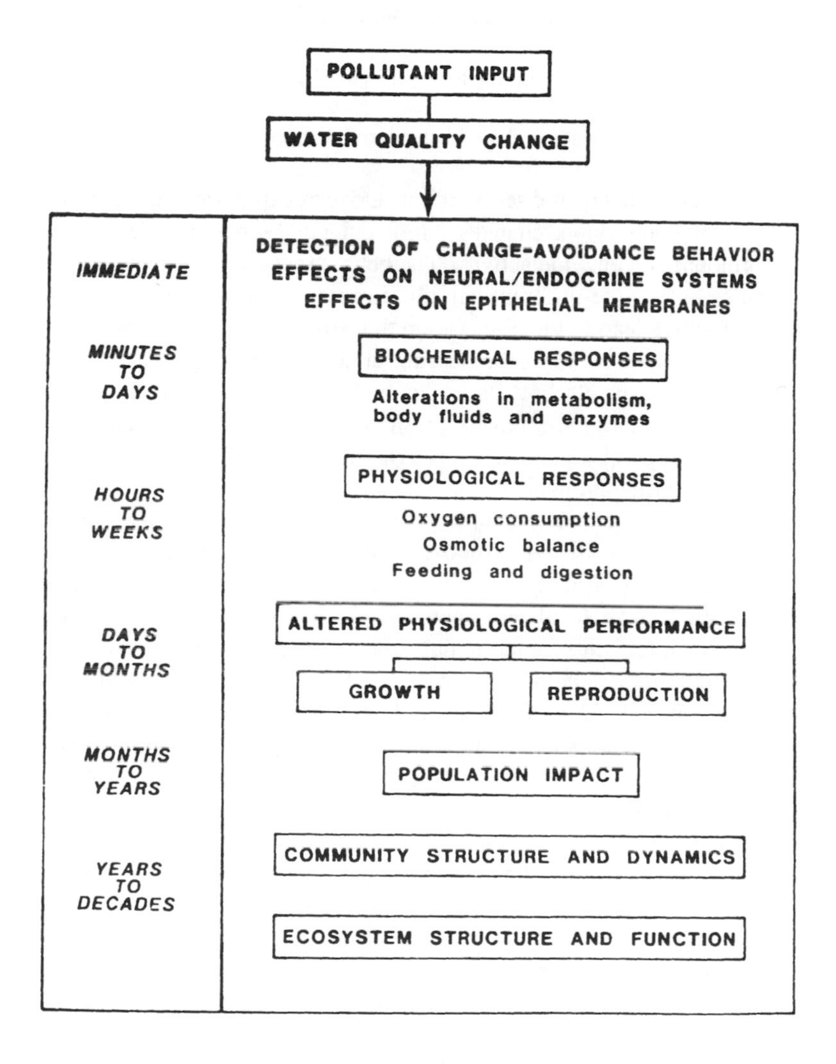

FIGURE 5. Temporal sequences of stress effects. (Modified from Sastry, A.N. and D.C. Miller,[5] Academic Press, New York, 1981, p. 265.)

Perspectives on stress will vary with the background of the observer. To the pathologist many of the responses to stressors can be described as or result in "disease," if disease is defined as "any departure from normal structure or function of the animal." Specific responses to stressors, from the viewpoint of the pathologist, include (in probable descending order of priority):

1. The classical cell and tissue changes such as *inflammation* (acute and chronic), *degeneration* (including edema, necrosis, and metaplasia), *repair and regeneration* (proliferation, hyperplasia, and scar formation), *neoplasia* (including consideration of cell origin, stage, and type — whether benign or malignant), and *genetic derangement* (including chromosomal changes and some skeletal abnormalities)
2. Changes in resistance to infection — usually by reduction in immunocompetence
3. Physiological, biochemical, and behavioral changes

The biochemist would see all of the above as expressions of altered cellular metabolism that produce changes at molecular and subcellular levels, mediated by hormonal and enzyme activity. The behaviorist and the physiologist would of course have quite different but still correct perceptions of stress responses — all of which lead to the observation that stress from environmental changes can be an excellent, all-encompassing concept in biology, but care must be taken not to allow its boundaries to become so broad that it loses utility. Attempts to quantify the effects of stress must persist, as must attempts to determine the precise pathways through which stress from pollution exerts its impact on marine animals.

There is another small complication to this already complex story of stress responses. The *nonspecific* responses just described as the general adaptation syndrome and illustrated in Figure 3 have to be distinguished from *specific* stress responses — localized reactions, physiological and/or morphological, to injury or infection of a particular tissue or organ. These are superimposed on and may modify the nonspecific reactions to the same stressor.

Stress in individual fish can and will be reflected in changes in fish populations — either short-term or long-term, ranging from dramatic to subtle — affecting survival, growth, and reproduction, and hence expansion or diminution of the population. The extent of population response (as a reflection of collective individual responses) will depend on the intensity and duration of environmental change. Each species has a series of physiological life zones with respect to variations in any stressor. The animal usually functions in a zone of normal adjustment and has a limit of compensation for changes in any environmental factor, as shown in Figure 6. Beyond this limit, the animal functions with increasing energy expenditure, and disabilities appear — some reversible if the environmental change is not too severe or too prolonged, and others irreversible and fatal if the change is drastic or prolonged. This diagram is worthy of some attention; it encompasses the entire range of pollution effects on individuals.

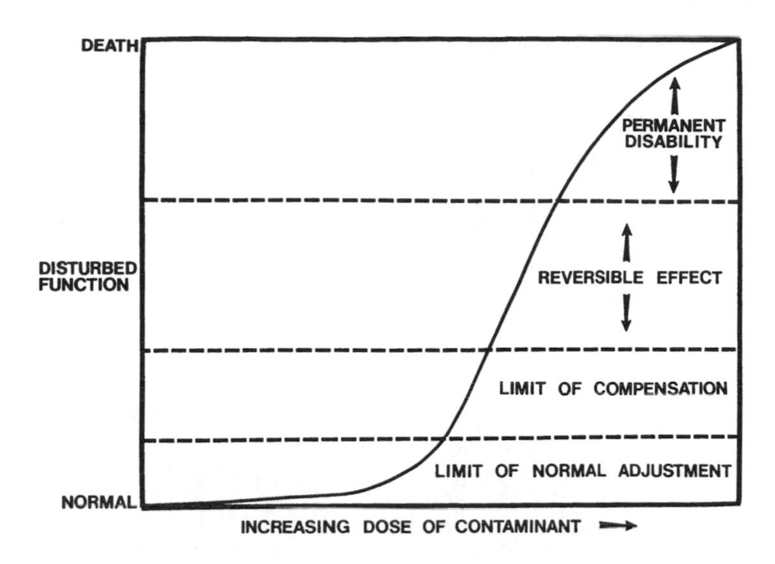

FIGURE 6. General life zones in the presence of a varying environmental factor. (Modified from Wilson, K.W.,[6] *Rapp. P.-V. Reun. Cons. Int. Explor. Mer.* 179: 333, 1980.)

The literature on stress in fish is abundant; books have been written on the subject.[7] Although most studies of the physiological and morphological consequences of stress have emphasized the vertebrates, and particularly humans, there are indications that counterpart phenomena may exist in the lower animals as well — and we should definitely learn more about these. In bivalve molluscs, signs of stress may include mantle recession, pale digestive gland, regression of digestive tubule epithelium, hemocyte infiltration of tissues, edema, lag in gametogenesis, shell abnormalities, and increased ceroid (brown bodies) — all or most of which constitute a stress syndrome, as shown in Figure 7. In the larger crustaceans, signs of stress include black gills, abdominal muscle opacity, molt retardation, exoskeletal overgrowth with filamentous bacteria and protozoan epibionts, frequent occurrence of shell disease, disoriented or inappropriate behavior, presence in the tissues of gram-positive bacteria, and clotting of the hemolymph (as a response to gram-negative bacterial endotoxin) — again constituting a stress syndrome, as shown in Figure 8.

As a final point in this discussion of stress in marine animals, we should note the emerging body of information about the protective role of *stress proteins* as part of the adaptive response of vertebrate and invertebrate organisms to potentially harmful environmental changes — chemical, physical, and biological. Exposure to a stressor results in the rapid synthesis of these stress proteins, which are in the range of 60 to 70 and 80 to 90 kDa, and the suppression of synthesis of other proteins. Although the cellular mechanisms of induction and activity of these proteins are not fully understood, the result is an increase in tolerance of the animal, even to other types of stressors. The

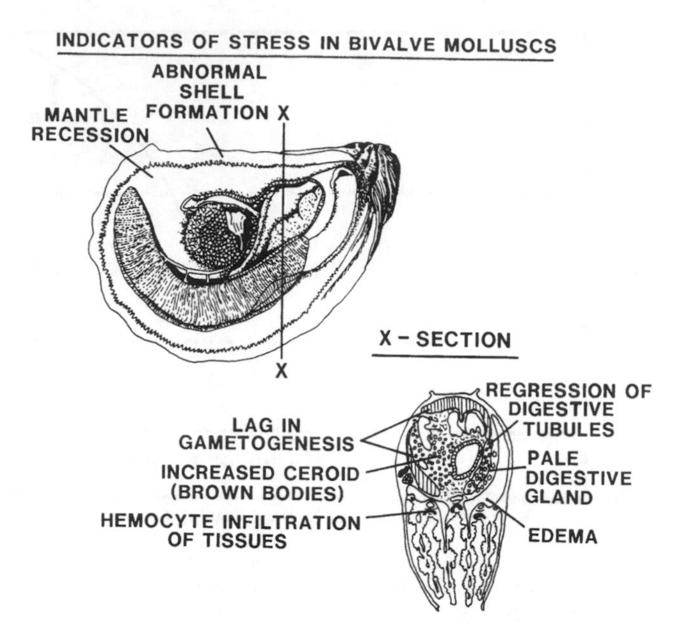

FIGURE 7. Some responses of bivalve molluscs to stressors.

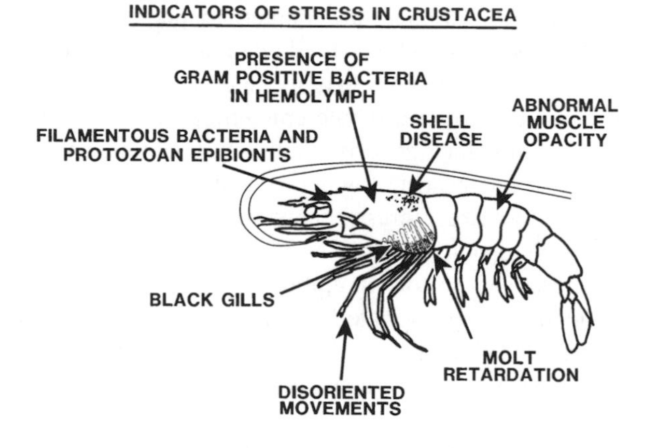

FIGURE 8. Some responses of crustaceans to stressors.

stress proteins thus represent a significant expansion of the concept of a stress response, beyond the physiological/biochemical changes that constitute the generally understood adaptation syndrome in fish and the counterpart adaptive responses in invertebrates (stress proteins can be induced in invertebrates as well as in the vertebrates).

So, with this long and complicated but essential discussion of stress as a foundation, we should now be ready to move to a second, but closely interconnected, concept in pollution biology — the physiological/biochemical mechanisms that lead to adaptation and survival or disability and death in degraded habitats.

REFERENCES

1. **Selye, H.** 1952. *The Story of the Adaptation Syndrome*. Acta, Montreal, Quebec, 255 pp.; **Selye, H.** 1955. Stress and disease. *Science* 122: 625–631.
2. **Brett, J. R.** 1958. Implications and assessments of environmental stress, pp. 69–97. in *The Investigation of Fish Problems,* H. R. MacMillan Lectures in Fisheries, University of British Columbia.
3. **Esch, G. W., J. W. Gibbons, and J. E. Bourque.** 1975. An analysis of the relationship between stress and parasitism. *Am. Midl. Nat.* 93: 339–353.
4. **Mazeaud, M. M., F. Mazeaud, and E. M. Donaldson.** 1977. Primary and secondary effects of stress in fish:; some new data with a general review. *Trans. Am. Fish. Soc.* 106: 201–212; **Peters, G., H. Delvanthal, and H. Klinger.** 1980. Physiological and morphological effects of social stress in the eel (*Anguilla anguilla* L.). *Arch. Fisch. Wiss.* 30: 157–180.
5. **Sastry, A. N. and D. C. Miller.** 1981. Application of biochemical and physiological responses to water quality monitoring, pp. 265–294. in Vernberg, J., A. Calabrese, E. P. Thurberg, and W. B. Vernberg (Eds.), *Biological Monitoring of Marine Pollutants*. Academic Press, New York. 559 pp.
6. **Wilson, K. W.** 1980. Monitoring dose-response relationships. *Rapp. P.-V. Reun. Cons. Int. Explor. Mer* 179: 333–338.
7. **Wedemeyer, G. A., F. P. Meyer, and L. Smith.** 1976. *Environmental Stress and Fish Diseases*. T.F.H. Publishers, Neptune, N.J. 200 pp.; **Pickering, A.,** (Ed.). 1981. *Stress and Fish,* Academic Press, London. 365 pp.; **Bayne, B. L., D. A. Brown, K. Burns, D. R. Dixon, A. Ivanovici, D. R. Livingstone, D. M. Lowe, M. N. Moore, A. R. D. Stebbing, and J. Widdows** (Eds.) 1985. *The Effects of Stress and Pollution on Marine Animals*. Praeger Scientific, New York. 384 pp.; **Adams, S. M.** (Ed.) 1990. *Biological Indicators of Stress in Fish*. Symposium 8, American Fisheries Society, Bethesda, MD, 191 pp.

2 Adaptation and Survival or Disability and Death in Polluted Habitats

****

PACIFIC OYSTERS ON THE COAST OF FRANCE

Pacific oysters, Crassostrea gigas, were introduced to the coastal waters of France by mass importations beginning in the late 1960s, after the native oysters had drastically declined in abundance, due mostly to effects of epizootic diseases. Millions of seed oysters were airlifted from Japan during the period 1968 to 1974, with high hopes of reestablishing the industry in such traditional French oyster growing areas as Arcachon, Oleron, Marennes, and La Trinité. Initial results of the mass transplantation were encouraging. The introduced species survived, grew, and even reproduced in some protected bays. By the mid-1970s though, indications of a severe problem were appearing in some of the bays. The new oysters were exhibiting poor growth and grossly malformed shells. Shell abnormalities reached an intolerable level of 90 percent in the Bay of Arcachon in 1980–1982.

Crisis response research conducted during the late 1970s and early 1980s in France and Britain eventually disclosed the cause of the deformities as an environmental pollutant — tributyltin — an organic compound being used as an ingredient in antifouling paint for small boats, that was leaching out in growing areas. The problem, which was for several years a real threat to the successful reestablishment of the oyster industry in France, was solved with the immediate imposition of a ban on the use of organotin compounds in antifouling paint for boats. Prevalences of the shell abnormalities fell to negligible levels soon after the ban took effect, and the oyster industry regained momentum in the growing areas that had been affected.

One of the fascinating aspects of this scientific detective story is the extreme sensitivity of the Pacific oyster to even minute concentrations of the specific environmental contaminant. The research demonstrated clearly that the presence of this single toxic chemical, in vanishingly small concentrations, could have a striking effect on the physiology of shell deposition in the oyster (probably by disrupting normal calcium metabolism), and ultimately on the marketability of the product. A more ominous finding was that those extremely low concentrations of tributyltin could also kill oyster and crab larvae — pointing to potential impacts on population abundance. (Later studies showed that juvenile crabs were also very sensitive to the contaminant; experimental exposures to tributyltin retarded limb regeneration, delayed molting, and produced deformities in regenerated appendages.)

The tributyltin/oyster episode is only one of many examples of disabilities and deformities that can be attributed to chemical contamination of inshore habitats. Too often, though, such a clear cause and effect relationship of abnormalities with specific contaminants has not been demonstrated — being obscured by the simultaneous presence of other suspect pollutants.

From "Field Notes of a Pollution Watcher"
(C. J. Sindermann, 1986)

****

INTRODUCTION

As Rachel Carson has described so well in her book *The Edge of the Sea,* the margins of the oceans are transition zones between land and sea, and are therefore places that demand every shred of adaptability that living things can muster. Humans have increased those demands, sometimes to intolerable levels, by adding toxic chemicals or excesses of nutrient chemicals to inshore waters. We have, for instance, added pesticides and other synthetic industrial chemicals that can, even in low concentrations, drastically affect the physiology of fish and shellfish (and even aquatic plants), and with which the species may have had no previous evolutionary experience. Heavy organic loads, in the form of sewage sludge and effluents, have been superimposed on some aquatic habitats, and they can produce anaerobic or low-oxygen environments. Effluents often contain contaminants such as heavy metals and chlorinated hydrocarbons that interfere with the physiology and biochemistry of the fish and the food organisms that they consume.

There has been a gradual increase in the extent of polluted coastal/estuarine waters — an increase that is generally proportional to the density of the adjacent human population and its level of industrialization. Excellent documented examples of this exist in North America; counterparts exist in Europe and elsewhere in developed and developing countries.

Pollutant-induced stress in fish has been demonstrated repeatedly, and descriptions of lethal and sublethal effects of toxic trace metals, petroleum compounds, and halogenated hydrocarbons abound in the experimental literature. Pollutant stressors can exert lethal or sublethal influences at any point in the life cycles of fish or their food organisms. Figure 9 illustrates the life cycle of a bottom-dwelling species, the winter flounder, *Pleuronectes americanus*. It seems almost predictable that with so many potential impact points throughout life cycles, populations of fish in contaminated waters should decline and disappear, yet our experience on the east coast of North America indicates that this has not happened. Only in grossly contaminated coastal/estuarine waters do we see disappearances of some species. Even in such areas other species, which should have been affected, are still present, and in some instances their local abundance has increased during the past several decades. This is true particularly of some of the coastal/estuarine species that spend much of their lives in waters that are to some extent contaminated — especially the nursery areas for early life history stages.

A brief consideration of the facts will lead to the logical conclusion that death and disappearance are too harsh as criteria for determining whether pollutants are harmful to marine biota or not. Mortality is a questionable criterion for assessing the effects of pollution on fisheries — since it is a factor with low sensitivity and insufficient early warning capacity, and since reliable data on fish abundance are difficult to acquire, even in the absence of pollution. We should look instead at *sublethal effects* of pollutant stress as indicators of long-term impacts on populations that as yet are difficult to quantify and difficult to sort out from the effects of the other environmental stressors that were listed in Chapter 1.

Sublethal effects of pollutants may take many forms — physiological, biochemical, behavioral, and pathological. Physiological abnormalities include reduction or inhibition of reproductive capacity, growth retardation, and immunosuppression. Examples of biochemical disturbances include heavy metal effects on cellular enzymes, organochlorine interference with calcium metabolism, inhibition of amino acid absorption in the presence of heavy metals, contaminant influences on lysosomal stability, and a host of other effects. Behavioral changes include reduced chemosensory capacity, modification of migratory patterns, and increased vulnerability to predators. Among the numerous pathological signs are fin erosion, ulcerations, liver tumors, and skeletal anomalies, as illustrated in Figure 10.

Two sublethal effects of pollution deserve particular discussion:

- *Effects on early life history stages,* which are fortunately easier to demonstrate than effects on abundance (included here would be mutagenic and cytotoxic effects on embryos, and counterpart studies of gross abnormalities in fish larvae as a consequence of exposure to contaminants). The entire process of reproduction

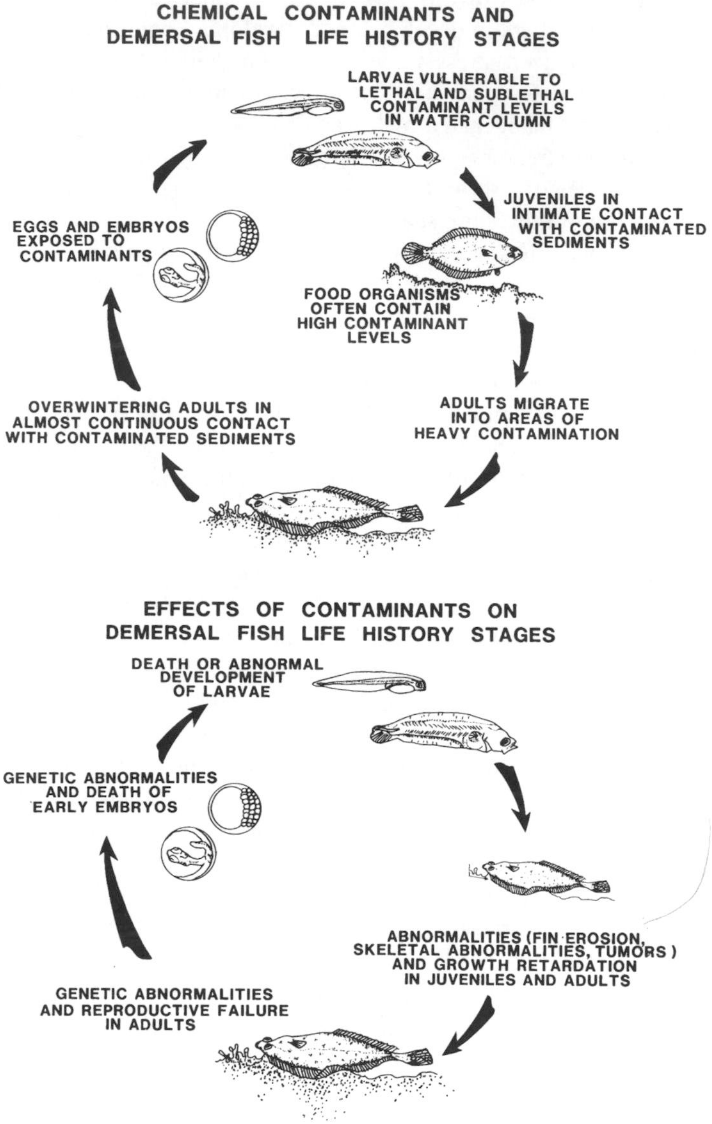

FIGURE 9. Life cycle of the winter flounder, *Pleuronectes americanus*, with potential pollutant impact points *(above)* and the effects of pollutants *(below)*.

can be affected by pollutants, beginning with modification of hormonal control of egg development and production (for example, through increased breakdown of estrogen by cytochrome oxidases induced by organic contaminants) and extending through effects on early growth of embryos and larvae, as affected by enzyme alterations.

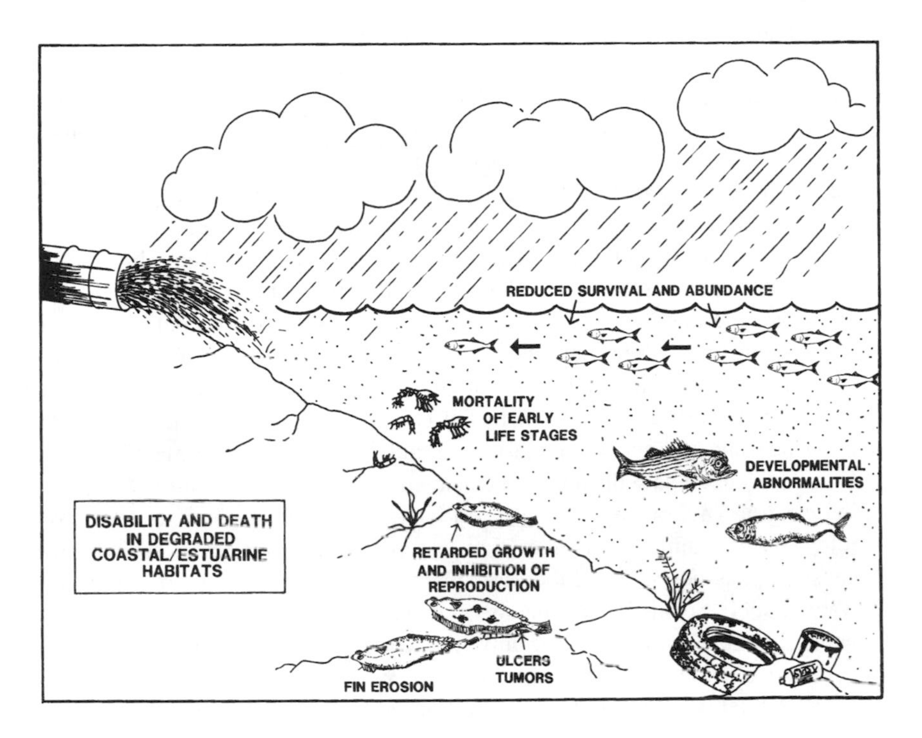

FIGURE 10. Some examples of the effects of contaminants on marine organisms.

- *Modification of enzyme activity* within the cells, resulting in disturbed metabolism, which would be reflected at higher levels of organization. An excellent example of this would be sublethal effects of metal exposure on cellular enzymes. Such effects include energy-requiring chronic demand for compensatory induction of enzymes or blocking of sensitivities by which enzyme reaction rates are regulated. This lessens the metabolic flexibility necessary for an animal's adaptation and survival during environmental challenge. Another example is seen in induction of so-called mixed function oxygenases (to be described later in this chapter) by chlorinated hydrocarbons; such induced enzymes have been implicated in disturbances of reproductive physiology, probably by altering liver steroid metabolism.

It is important to note at this point, however, that the extent of harm inflicted on the oceans and their inhabitants by humans and their activities can be overestimated (and conversely, the survival potential of marine animals in degraded habitats can be underestimated). It is easy to view with alarm and to predict dire consequences when surrounded by severe but localized aquatic environmental degradation, as in the immediate vicinity of major population centers such as New York City. It is not as easy to project such consequences very far beyond the impacted zones (which may, however, be increasing in size and in degree of degradation as a consequence of ocean currents, airborne fallout, and increased human population density). Not all responses of animals to contaminants are disruptive, and a number of adaptive strategies exist that can mediate pollution impacts.[1] The subject of abnormality and death in degraded habitats is always of great popular interest and contains elements of reality, but *survival* in degraded habitats is also a reality, so mitigating aspects must not be ignored. We should feel some small obligation to look at both sides of the marine pollution mirror — the dark side of disability and death in degraded habitats and the shiny side of survival in those same habitats.

Natural processes — wave action, sedimentation, ocean currents, and chemical alteration — rapidly reduce the effects of many contaminants or dilute them below the point where they kill living organisms. Sublethal long-term effects may still be a danger, of course, but even these effects may occur in very restricted areas (except in the case of migratory species). Also, the resilience and tolerances of marine animals can be underestimated. All life forms are endowed with innate mechanisms to permit survival of the species in changing environments; some are illustrated in Figure 11. Fish have physiological/biochemical resources that lessen the effects of even such foreign substances as pesticides. Other human pollutant contributions to the planet's waters, such as oil, heavy metals, and nutrients, are substances that occur naturally and become harmful only in excess, which may still be only a few parts per million. Table 1 summarizes some important mechanisms of resistance or tolerance to pollutants, including (a) those mitigating factors that are *external* to the organism, (b) those that are *internal,* and (c) those that exist at a *population level*. Each of these will be examined in the following sections.

EXTERNAL MITIGATING FACTORS

Any consideration of coastal/estuarine contamination and its effects on fish must include the concept of *environmental filters* — which mitigate the impact of pollutants.[2] Estuaries are environmental filters that protect coastal fish populations from extremes of toxicity (except when the system is short-circuited by freshets, airborne contaminant movement, point discharges, or ocean dumping). Filtration is a consequence of physical, chemical, and biological activities, including adsorption of contaminants to colloidal and other particulates in the water column, sedimentation of particulates, uptake and metabolism of contaminants by noncommercial benthic and planktonic organisms; chemical modification of some contaminants by microorganisms; utilization of nitrates and phosphates from human wastes in primary production and plant growth; precipitation of some dissolved industrial contaminants; chemical degradation of some contaminants (petroleum, for example); and many others. The estuary thus can function as a buffer, reducing the concentrations; hence the toxicity and effects of anthropogenic chemical additions to coastal and oceanic waters. It is in this estuarine filter, however, where exacerbation of pollutant effects can be best seen — in physiological and pathological

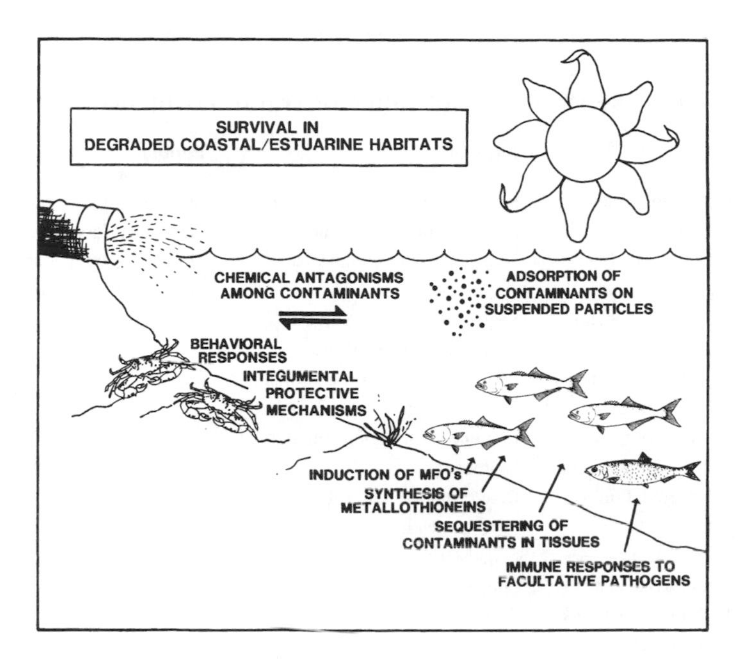

FIGURE 11. Some mechanisms that enhance the survival of marine organisms in degraded habitats.

changes in indicator organisms. Also in this buffering system will be seen the true range of responses of organisms — genetic, biochemical, physiological, and behavioral. Perhaps most important, many fish and shellfish use the estuaries during all or part of their life cycles.

An additional consideration in assessing the effects of pollutants on fish populations is *the relatively limited areal extent and small number of acutely contaminated zones* in coastal/estuarine waters, but the far greater extent of *marginally toxic zones* where sublethal (chronic) effects may occur. Acutely toxic areas include river mouths, toxic dump sites, the immediate vicinity of ocean outfalls, industrial outfalls, and sites of accidental toxic spills or deliberate discharges. Marginally toxic areas are associated with plumes extending from the acute zones in directions and distances dependent on the hydrographic features of that area.[3] In both instances, however, many life history stages of key oceanic and shelf species are dependent on these contaminated zones.

Prominent among the external mitigating influences are adsorption of contaminants on suspended particles, with subsequent removal from the water column by sedimentation and, more directly, by *antagonisms* of heavy metals, reducing the net effect on physiological processes.[4] Considering the complex mixture of toxic trace metals in most polluted zones, such antagonisms may be common, sharply reducing the net effects of the contaminants. (It should be kept in mind, however, that the direct opposite effect — potentiation or synergism — can occur with certain heavy metal combinations.)

TABLE 1
Mechanisms of Resistance or Tolerance of Fish to Pollutants

External	Internal	Population
Environmental mitigation Chemical antagonisms (for example, metals — reducing the additive effects on physiological processes) Adsorption of contaminants on suspended particles and removal from water column by sedimentation Microbial degradation of organics Behavioral modification Retreat from toxic zones Altered migration patterns Integumental protective mechanisms Gill mucus Intestinal mucus Body surface mucus layer Antibodies in mucus	Synthesis of metallothioneins Induction of mixed function oxygenases to metabolize organics Sequestering of fat-soluble toxic substances in fat Shift to lactic acid metabolism in low-oxygen environment Immune responses to facultative pathogens Rate of uptake and release of heavy metal species may vary Sequestering of contaminant in shell (Crustacea)	Selection by differential mortality for efficiency in coping with contaminants (for example, high antibody responders)

INTERNAL MITIGATING FACTORS

Responses of aquatic animals to a changing chemical environment as a consequence of pollutants may take a number of forms. A generalization that is becoming increasingly apparent, but is often overlooked, is that *aquatic organisms, populations, and ecosystems are equipped with a wide variety of homeostatic or resistance mechanisms that permit survival in the presence of certain pollutants*. Individuals can tolerate or at least survive levels of contaminants that are within physiological limits, and tolerances may increase with continued exposure to sublethal doses of the contaminant[5] (in some instances it may decrease, however). Furthermore, individual animals may respond to organic contaminants by chemically or physically sequestering them, or by the induction of enzymes that metabolize the foreign chemicals. Detoxification of organic chemical pollutants through a number of metabolic pathways can be effective, although there are instances in which the transformed (metabolized) compound (for example, benzo[a]pyrene) can be more toxic or carcinogenic for the animal than the compound itself. Trace metal toxicity can be reduced by protein binding. Some specific methods of adaptation to environmental pollutants include the following.

Heavy Metal "Traps"

Among the internal mechanisms mitigating the effects of heavy metals, recent studies[6] have demonstrated the presence in marine fish of mucus complexes with high copper-, zinc-, and cadmium-binding capacity. The phenomenon is partly physical-chemical and partly an extension of biological manipulation of metal salts. In the normal environment, fish swallow sea water, and, aided by the pH of the intestinal contents, calcium and magnesium are precipitated out in mucus strands. In waters polluted by cadmium and other heavy metals, high concentrations of those metals are also precipitated out, and the granules of metal salts are incorporated into mucus complexes that are subsequently eliminated. With this process, levels of heavy metals in the intestinal lumen do not become excessive, and thus are not absorbed by the fish. The mechanism, demonstrated in eels, *Anguilla anguilla,* and other species, consists of mucus secretion in the anterior intestine, creating extracellular "mucus traps" in which the heavy metal precipitates are incorporated and eliminated with the feces.

Another extracellular trap for heavy metals consists of increased mucus production by the gills in the presence of metal intoxication. Experimental studies have demonstrated fixation of mercury by gill mucus and its subsequent elimination, thus preventing high metal concentrations from contacting the gill epithelium.

An "intracellular trap" for heavy metals consists of binding of the metals (such as cadmium and mercury) to low-molecular-weight proteins with a high content of the amino acid cysteine (metallothioneins). The presence of heavy metals induces biosynthesis of *metallothioneins* in tissues; the bound metal is toxicologically inert, providing tolerance to high contamination levels in chronic exposures (the metallothioneins appear when the high-molecular-weight soluble proteins reach saturation with cadmium or certain other metals).

Induction of Mixed Function Oxygenases

A number of organic pollutants are known to induce so-called "mixed function oxygenases" (MFOs) in fish. These are enzymes that participate in metabolism and degradation of several categories of foreign compounds in the animal.[7] Oxidized metabolites of toxic foreign organic compounds can be eliminated by diffusion across membranes or they can be conjugated with serum components and then excreted. The toxic compounds are eventually metabolized to less toxic ones and excreted, although as noted earlier in this chapter there are examples of metabolites being more toxic than the parent compound.[8] Some compounds, such as the aromatic hydrocarbons, act as strong inducers of MFOs, whereas others, such as certain of the polychlorinated biphenyls, are poor inducers. The MFO system is particularly useful as a mechanism for detoxifying short-term sublethal doses of organic contaminants, but may not be as effective in ensuring long-term survival in chronically or heavily polluted habitats.[9]

Induction of other kinds of physiological changes that increase resistance in fish has also been demonstrated with several organochlorine and petroleum compounds. Mechanisms of resistance to two pesticides in one study included a *membrane barrier* that reduced uptake of the contaminants and a *brain barrier* in the form of insensitivity at the target site.[10]

Immune Responses

Although suppression of immune responses is often cited as one of the important responses of fish to exposure to pollutants, there are aspects of internal resistance that tend to favor survival of fish in polluted habitats. In one experiment, exposure of cunner, *Tautogolabrus adspersus,* a small inshore species, to cadmium did not reduce humoral antibody production but did reduce the bactericidal capabilities of phagocytes.[11]

In a related field study, a survey of antibodies to a wide spectrum of bacteria in summer flounder, *Paralichthys dentatus,* from polluted and unpolluted habitats found significantly higher antibody levels and a greater diversity of antibodies in samples from polluted waters (the New York Bight).[12] Weakfish, *Cynoscion regalis,* also exhibited increased titers against many bacteria. The greatest proportion of increased titers was against *Vibrio* spp., although prominent titers against other fish pathogens were seen.

In a subsequent experimental study with summer flounder, a greater proportion of high-antibody responders was found in fish taken from polluted waters than in those from unpolluted sites.[13] Fish pathogens, *Vibrio anguillarum* and *Aeromonas salmonicida,* elicited particularly strong responses. The investigators suggested that fish surviving in polluted areas may be genetically selected as high-antibody responders. Other recent studies have demonstrated that many organic contaminants (for example, dioxins, PCBs) are immunosuppressive and that some metals may selectively suppress critical parts of the immune system.

MITIGATING FACTORS AT THE POPULATION LEVEL

On a population level, there may be long-term adaptations to high environmental levels of naturally occurring contaminants such as heavy metals and petroleum hydrocarbons. Part of the process is the selection of resistant strains, which is feasible since many species have a high reproductive potential and high genetic plasticity. Some evidence exists for selective action of pollutants (in addition to selection for increased immunological competence). In one study of killifish *(Fundulus heteroclitus),* some females from unpolluted coastal areas were found to produce eggs that were much more resistant to methylmercury than eggs from other females (as measured by percentages of developmental anomalies that followed exposure).[14] When a population from a heavily polluted coastal area was examined, a much higher percentage of the females produced "resistant" eggs. However, more recent studies by the same investigators indicated that even though embryos from polluted areas were more resistant to methylmercury toxicity, adults seemed less tolerant, as determined by mortality and rate of fin regeneration.[15]

A number of separate lines of investigation converge to illustrate other strategies employed by fish and invertebrates for survival in degraded habitats. Examination of "scope for growth" — the relative energy budget (energy available for growth and gamete production) of animals in polluted zones as compared to unpolluted zones — demonstrated that at a critical level of contamination animals were living at an energy deficit. Part of the survival strategy for species such as mussels was to reduce reproductive output during periods of high contamination, thereby conserving energy for growth.[16] Population replacement during such periods would depend on recruitment from populations outside the deficit zone — a good strategy so long as contamination is not uniformly or extensively distributed. Resistance strategies impose a burden on the animal in that they are energy requiring, placing it at a physiological and perhaps survival disadvantage relative to other members of the species living in nonpolluted zones.

This kind of perspective on pollution effects suggests that *damage occurs as a consequence of exposure to pollutants, but that important mechanisms for survival in modified habitats may be mobilized to mitigate part of the damage* — although at the expense of energy (Figure 12). What emerges is, almost predictably, a limited system of checks and balances. Physiological mechanisms that compensate for environmental chemical imbalances function up to a point; beyond such limits the phenomena of toxic effects, collapse, and death appear; sequestering, detoxifying, and excreting toxic chemicals are forms of protection that may be overloaded or overwhelmed. Examples of protective mechanisms include induced MFOs to reduce toxicity of synthetic organics, protein binding and mucus traps to reduce heavy metal toxicity, and the possible selection of high-antibody responders as compensation for the immunosuppressive effects of some pollutants. One significant "downside" to these mechanisms may be, however, that survivors may build up high levels of contaminants in their tissues, thus posing chemical threats to predators, including human consumers.

This chapter and the one preceding it have explored two interrelated concepts in pollution biology: the role of pollutants as *stressors* for marine fish and shellfish and the *physiological/biochemical mechanisms* that enable survival in degraded habitats, if the effects of the stressor are not too severe or too prolonged. For those readers who have managed to survive the necessarily technical treatment of these concepts, a number of conclusions should have emerged. Some that seem supportable by the available data include these:

DISABILITY AND DEATH IN DEGRADED ESTUARINE/COASTAL HABITATS	SURVIVAL IN DEGRADED ESTUARINE/COASTAL HABITATS
REDUCED SURVIVAL AND ABUNDANCE RETARDED GROWTH INHIBITION OF REPRODUCTION MORTALITY OF EARLY LIFE STAGES DEVELOPMENTAL ABNORMALITIES POLLUTION-RELATED PATHOLOGY	EXTERNAL MITIGATION: DILUTION CHEMICAL ANTAGONISMS ADSORPTION OF CONTAMINANTS ON SUSPENDED PARTICLES BEHAVIORAL RESPONSES INTEGUMENTAL PROTECTIVE MECHANISMS INTERNAL MITIGATION: INDUCTION OF MFO's SYNTHESIS OF METALLOTHIONEINS SEQUESTERING OF CONTAMINANTS IN TISSUES IMMUNE RESPONSE TO PATHOGENS

FIGURE 12. A balance sheet for disability and death versus survival in degraded habitats.

- The fact that animals can be injured or killed by exposure to chemical pollutants is clearly supported by thousands of experimental studies.
- Environmental contaminant levels in a number of restricted coastal/estuarine areas are sufficiently high to produce the acute or chronic effects seen in experimental systems.
- Contaminant levels in tissues of fish have been examined in a number of geographic areas. In many instances, widespread occurrence of detectable levels of selected contaminants has been demonstrated (PCBs, DDT, mercury, petroleum components).
- In certain localized areas, tissue levels of contaminants have been recognized in fish that are well above the few legal action levels that exist; toxic effects on the fish of such levels are largely unknown, as are toxic effects of exposure prior to buildup of observed tissue levels. Beyond this, the relationship between toxicity and any tissue level has been inadequately explored. High body burdens of contaminants may be sequestered and not affect health, until a level is reached at which accommodation is no longer feasible and spillover may occur.
- Inadequate data are available about synergistic effects on fish of extremely complex mixtures of contaminant chemicals and dissolved or particulate organics in polluted waters.
- Enough experimental data exist to state that antagonisms and synergisms in complex mixtures of contaminant chemicals may be important factors in producing any net effect.
- Some experimental data exist about differential effects of various species, isomers, and substituted forms of contaminant chemicals on fish, but the influences of chelation and degradation on biological activity must also be considered.
- An obvious need exists to examine pollution effects on fish within the broader context of other environmental variables that may influence survival and abundance. Within this context, more information is needed about the extent of coastal/estuarine waters where fish populations are actually at risk from toxic levels of contaminants. Corollary efforts are needed to quantify pollution effects

on all life stages of fish, at acute and subacute exposures to single and multiple contaminants, and in field as well as experimental situations.

- Risk analysis for resource species in any geographic area must take into account all the complex variables, natural and man-induced, that may affect survival.

Examining the pluses and minuses — the potentially harmful effects of pollutants on fish populations, as disclosed by experimentation, and also the possible mitigating factors (many of them also indicated by experimentation) — evidence at hand suggests damage to fish populations in areas of highest contamination in the form of lethal or sublethal effects, but little direct indication of widespread, overriding effects on fish stocks. Baseline data on possible population fluctuations in the *absence* of pollution are, however, still sufficiently inadequate to make most conclusions tenuous.

Polluting chemicals are being added in ever-increasing variety to coastal/estuarine waters, as effluents from inventive and seemingly insatiable human technologies. Fish and shellfish in affected habitats either live, survive marginally, or die, depending on their ability to adapt to the changed chemical environment. The methods of adaptation are marvelously varied; some principal physiological devices include inducing cellular enzymes to metabolize the foreign chemical; blocking the entrance of the chemical at the cell membrane level; creating mucus traps in the digestive tract for excess metals; sequestering fat-soluble chemicals in fat cells; and linking metals with proteins to reduce their toxicity. Other strategies to counter toxic chemicals include energy conservation mechanisms, especially reduced reproductive capacities, and exclusionary mechanisms such as increased mucus production in fish and prolonged shell closure in bivalve molluscs.

REFERENCES

1. **Capuzzo, J.** 1981. Predicting pollution effects in the marine environment. *Oceanus* 24(1): 25–33.
2. **Verduin, J.** 1969. Man's influence on Lake Erie. *Ohio J. Sci.* 69(2): 65–70.
3. **Munday, J. C., Jr. and M. S. Fedosh.** 1981. Chesapeake Bay plume dynamics from LANDSAT, pp. 79–92. in Campbell, J.W. and J.P. Thomas (Eds.), *Chesapeake Bay Plume Study: Superflex 1980*. NASA CP-2188 and NOAA/NEMP III 81 ABCDFG 0042.
4. **Weis, P. and J. S. Weis.** 1979. Congenital abnormalities in estuarine fishes produced by environmental contaminants, pp. 94–105. in *Animals as Monitors of Environmental Pollutants. National Academy of Science,* Washington, D.C.; **Weis, P. and J. S. Weis.** 1980. Effect of zinc on fin regeneration in the mummichog, *Fundulus heteroclitus,* and its interaction with methylmercury. *Fish. Bull.* 78: 163–166.
5. **Bryan, G. W.** 1976. Some aspects of heavy metal tolerance in aquatic organisms, pp. 7–34. in Lockwood, A. (Ed.), *Effects of Pollutants on Aquatic Organisms*. Cambridge University Press, New York; **Bryan, G. W. and L. G. Hummerstone.** 1971. Adaptation of the polychaete *Nereis diversicolor* to

estuarine sediments containing high concentrations of heavy metals. *J. Mar. Biol. Assoc. U.K.* 51: 845–863.

6. **Nöel-Lambot, F., J. M. Bouquegneau, and A. Disteche.** 1980. Some mechanisms promoting or limiting bioaccumulation in marine organisms. *Int. Counc. Explor. Sea,* Doc. C.M.1980/E:39. 25 pp.
7. **Payne, J. F. and W. R. Penrose.** 1975. Induction of aryl hydrocarbon (benzo[a]pyrene) hydroxylase in fish by petroleum. *Bull. Environ. Contam. Toxicol.* 14: 112–116; **Stegeman, J. J.** 1978. Influence of environmental contamination on cytochrome P450 mixed function oxygenases in fish: implications for recovery in the Wild Harbor marsh. *J. Fish. Res. Board Can.* 35: 668–674; **Stegeman, J. J. and D. J. Sabo.** 1976. Aspects of the effects of petroleum hydrocarbons on intermediary metabolism and xenobiotic metabolism in marine fish, pp. 423–436. in *Sources, Effects and Sinks of Hydrocarbons in the Aquatic Environment. American Institute of Biological Science,* Washington, D.C.
8. **Stegeman, J. J., T. R. Skopek, and W. G. Thilly.** 1979. Bioactivation of polynuclear aromatic hydrocarbons to cytotoxic and mutagenic products by marine fish. in *Carcinogenic Polynuclear Aromatic Hydrocarbons in the Marine Environment.* Ann Arbor Science Publishers, Ann Arbor, MI.
9. **Burns, K. A.** 1976. Microsomal mixed function oxidases in an estuarine fish, *Fundulus heteroclitus,* and their induction as a result of environmental contamination. *Comp. Biochem. Physiol.* 538: 443–446.
10. **Yarbrough, J. D. and M. R. Wells.** 1971. Vertebrate insecticide resistance: the in vitro endrin effect on succinic dehydrogenase activity on endrin-resistant and susceptible mosquitofish. *Bull. Environ. Contam. Toxicol.* 6: 171–177.
11. **Robohm, R. A. and M. F. Nitkowski.** 1974. Physiological response of the cunner, *Tautogolabrus adspersus,* to cadmium. IV. Effects on the immune system. U.S. Department of Commerce, NOAA Tech. Rep. NMFS SSRF-681, pp. 15–20.
12. **Robohm, R. A., C. Brown, and R. A. Murchelano.** 1979. Comparison of antibodies in marine fish from clean and polluted waters in the New York Bight: relative levels against 36 bacteria. *Appl. Environ. Microbiol.* 38: 248–257.
13. **Robohm, R. A. and D. S. Sparrow.** 1981. Evidence for genetic selection of high antibody responders in summer flounder *(Paralichthys dentatus)* from polluted areas, pp. 273–278. in *Proceedings of the International Symposium on Fish Biologics, Serodiagnostics, and Vaccines.* S. Karger, Basel.
14. **Weis, J. S. and P. Weis.** 1981. Methylmercury tolerance in killifish. *COPAS* 1(3): 35–36.
15. **Weis, J. S. and P. Weis.** 1987. Pollutants as developmental toxicants in aquatic organisms. *Environ. Health Persp.* 71: 77–85.
16. **Bayne, B. L., D. L. Holland, M. N. Moore, D. M. Lowe, and J. Widdows.** 1978. Further studies on the effects of stress in the adult on the eggs of *Mytilus edulis. J. Mar. Biol. Assoc. U.K.* 58: 825–841.

SECTION II:
Ocean Pollution and Disease

Chapter 1 of this book explored the role of stress in marine animals, especially the kinds of stress imposed by coastal pollution. Contained in that chapter was a borrowed diagram illustrating the concept that severe or continued stress resulted in increasing levels of disability and eventually in death. Disabilities are usually recognized and described as "disease", due to invasion by an infectious agent or to some structural or functional abnormality. Because pollution effects are so often expressed as disease, we need at this point in the discussion to develop some insights into the pathological processes involved and to look at examples of pollution-associated diseases in coastal/estuarine animals. Chapters 3 and 4 do these things, first for fish and then for shellfish.

We need also to look closely at another kind of pollution-related disease problem—the association of human diseases with ocean pollution. Here the choices are varied: some human diseases can be acquired by recreational contact with polluted waters (swimming and diving); some microbial diseases affecting humans may be acquired by eating raw or improperly cooked seafood taken from polluted waters; other microbial diseases may result from injuries such as shark bites or barnacle scrapes; and (far less commonly) some diseases may be acquired from long-term diets containing chemically contaminated seafood. Chapter 5 summarizes information about these public health aspects of ocean pollution, emphasizing the most severe problem—diseases resulting from eating microbially contaminated raw molluscan shellfish (oysters and clams).

Disease, then, is an important consequence of coastal pollution, whether looked at from a resource perspective (effects on fish and shellfish) or from a public health perspective (effects on humans). Pollution-associated diseases may reduce abundance of resource species or affect their growth and reproduction. Microbial or chemical contaminants accumulated by some marine animals may be transmitted passively to humans and may cause disease when the animals are eaten raw or are improperly processed. All of these intriguing possibilities will be explored in the three chapters that make up this section.

3 Ocean Pollution and Fish Diseases

****

THE FLOUNDER FROM BOSTON HARBOR

The winter flounder, technically labeled Pleuronectes americanus, lives and thrives in the coastal waters of the northeastern states, where it is a favorite catch of hook and line fishermen, and is also taken by inshore trawlers sailing from such historic ports as Gloucester, Rockland, New Bedford, and even Boston. On a cloudy morning in early June, Jim Driscoll, fishing off pier 21 in East Boston, brought up a specimen that ruined his day. It was clearly a winter flounder, but most of the fins had been eroded away, and near the tail were several large, bloody sores. Jim was tempted at first to throw the repulsive thing back, but the thought occurred to him that something must be very wrong down there to do such damage to any fish. He knew that there was a government statistical agent stationed down the street at the Custom House, so he wrapped the animal in a newspaper and delivered it, asking for some kind of explanation, but not really expecting much.

Jim wasn't aware of it, of course, but the statistical agent was part of a network of people with similar jobs deployed in all the major northeastern fishing ports,to collect data on catches and environmental conditions, and to transmit the information back to a research center in Woods Hole. One of the research staff members in that center actually looked at the specimen and even got excited about its condition — to the point where she called Jim a few weeks later with a report. The fish, she said, was suffering from three abnormalities, all associated with polluted habitats: the eroded fins and the skin ulcers were obvious, but additionally, microscopic examination of the liver disclosed numerous tumors — again characteristic of badly degraded habitats like Boston Harbor. Furthermore, chemical analyses of the flesh of the fish showed much higher than normal levels of PCBs, aromatic hydrocarbons and cadmium. She then gently warned him not to eat fish from the harbor too often.

Now Jim is no dummy. He knew that the water below the pier couldn't be quite the quality of a mountain spring, but he had not previously been confronted with visual evidence of just how dreadful it really must be. That single experience really upset him — to the extent that he gave up fishing altogether, and now doesn't even eat fish anymore, regardless of the source. He has concluded, not without cause, that the coastal pollution problem near big cities like Boston is probably intractable, and that seafood from those areas might even be dangerous to his health and that of his family.

From "Field Notes of a Pollution Watcher"
(C. J. Sindermann, 1988)

****

INTRODUCTION

During the past two decades, important new findings have added significantly to existing information about the interactions of pollution and disease in fish. Liver tumors and other lesions have been reported from flatfish sampled in grossly polluted estuarine locations in the United States and Europe. Genetic abnormalities in developing fish embryos and structural anomalies in larvae have been found in several studies to be related to the extent of chemical pollution. Fin erosion in fish has been postulated to be a consequence of continuous exposure of the skin to toxic levels of chemicals, combined with hormonal or metabolic stress-related disturbances within the animal. Ulcerations in fish have been recognized as a worldwide phenomenon, with diverse microbial etiologies, but with some tentative association with degraded habitats.

The overriding influence of stress-inducing environmental conditions in causing disease has become apparent in many recent studies and was examined in Chapters 1 and 2. The physiological/biochemical pathways through which long-term stress can result in pathology have also been described, so it is now time to examine in detail some of the consequences of pollution, as expressed by the occurrence of disease conditions in fish from degraded habitats.

The objectives of this chapter are: (1) to examine our present understanding of several specific disease conditions in fish that may be associated with habitat degradation, and (2) to assess the validity of the associations that have been proposed. Most of the disease conditions considered here are traceable to tissue damage and metabolic disturbances induced by toxic chemicals, or (in the case of infectious diseases) to increased infection pressure from expanded populations of facultative pathogens, combined with suppression of internal and external defense mechanisms in the stressed animals.

The evidence, though substantial, for an association between polluted habitats and fish diseases, should still be considered "circumstantial" or "inferential," since *direct cause-and-effect relationships have not been and may never be demonstrated to the satisfaction of all observers*. It seems nonetheless that some degree of association between certain fish diseases and pollution has been established. Studies employing a combination of field observations and experimental exposures to contaminants can do much to reduce uncertainties further, but the available body of evidence indicating a *statistical* relationship between degraded coastal/estuarine environments and certain disease conditions in fish has grown to a point where many observers have concluded that a relationship *does* exist, and that what remains to be done is to acquire additional quantitative data and to augment supporting experimental information — especially that concerning the physiological/biochemical bases for observed pathology. Additionally, much more difficult studies of the effects of pollution-associated diseases on the *abundance* of resource species are still necessary. Results of analyses of disease-caused effects on survival are important to resource managers, even though such investigations constitute only a subset of the scientific examination of pollution/disease relationships.

For a semblance of structure, the discussion in this chapter has been divided into two major sections: "pollution and infectious diseases" and "pollution and noninfectious diseases," with a minor niche for those diseases of temporarily "uncertain causation." The overriding principle that seems to be emerging is that *whether the cause of a disease is infectious or noninfectious, pollution stress can be a significant contributing factor in its occurrence in degraded habitats*.

POLLUTION AND INFECTIOUS DISEASES

In any attempt to consider diseases of marine fish within the larger context of coastal or estuarine pollution, it is important to reemphasize the role of stress — which was explored briefly in Chapter 1. A basic concept, which is being reinforced repeatedly, is an obvious one: most pollutants are chemical stressors, and much of what we call "pathology" or "disease" is a consequence of actions of environmental stressors. Many of the responses to extreme or prolonged stress can be described as or result in "disease" if disease is defined as "any departure from normal structure or function of the animal," or as "the end result of interaction between a noxious stimulus and a biological system." Furthermore, in attempts to define stress, it must be kept in mind that the infectious or noxious agent also acts as a stressor.

Dysfunction and death due to the activity of infectious agents constitute the narrower but often predominant perception of "disease." Infectious diseases — caused by viruses, bacteria, fungi, protozoa, and other parasites — are usually prime suspects in searches for the causes of fish mortalities, sometimes to the exclusion of other possible causes.

Infectious diseases usually exist at low levels in any population, weakening or disabling individuals and rendering them more susceptible to predators or other environmental stressors. Occasionally, though, disease outbreaks (epizootics) and mass mortalities comparable to the great human plagues of the Middle Ages may sweep through animal populations. In marine species, we have seen such massive epizootics at work in the great herring mortalities of the mid-1950s in the Gulf of St. Lawrence and the extensive oyster mortalities of the 1960s and 1980s in the Middle Atlantic states. These epizootics are triggered by a complex interplay of pathogen, environment, and host population. Considering only the environmental aspects of such outbreaks, any departure from normal conditions produces a degree of stress on individuals within a population and may contribute to an increase in the prevalence of a pathogen. Some of these environmental factors are abnormal temperatures or salinities, low dissolved oxygen, lack of adequate food, toxic chemicals, or high population densities. Resistance of the host animals to disease is of course intimately related.

For purposes of this discussion, infectious diseases include those resulting from invasions of pathogens and those resulting from activation of latent infections (already present in the host animal). Two infectious diseases that seem associated, at least in some studies, with polluted habitats are *lymphocystis*, a viral infection of the skin cells of fish, and *ulcers*, caused by a number of opportunistic microorganisms — viral, bacterial, and fungal — but especially by bacteria of the genus *Vibrio*. In these and other examples,environmental stressors can reduce resistance of the fish to infection by the microorganisms that are part of the normal flora of coastal/estuarine waters.

Activated Latent Infections: Lymphocystis

Lymphocystis is a viral disease that causes extreme enlargement of skin cells in many freshwater and marine fish (Figure 13) and has been postulated to be associated with environmental stressors. As an example, in a 1971 survey of the Irish Sea, three diseases — lymphocystis, epidermal ulcers, and fin erosion — were found to be abundant in the flatfishes, in plaice and dab.[1] Lymphocystis infection levels in fish taken in individual trawl catches ranged from 0 to 25% in plaice and from 0 to 17% in dab. The investigators pointed out that the Irish Sea had been used for dumping of toxic wastes, particularly PCBs, but their concluding statement was "...there is insufficient evidence to be certain whether the increased incidence of the diseases noted in 1971 is the result of an outbreak of epidemics of purely biological origin or if the dumping of toxic wastes is responsible."

An independent survey of lymphocystis in the Irish Sea in 1972 also found lymphocystis to be the most abundant of grossly visible pathological conditions, with highest prevalence (14.6%) in flounder, *Platichthys flesus*, and lesser prevalences in other flatfish (1.9% in plaice and 1.1% in dab).[2] Unlike the previous investigators, the new team considered recent pollution of the northeast Irish Sea to be the *least* likely explanation for high levels of lymphocystis,

FIGURE 13. Lymphocystis in striped bass.

pointing out that the disease had been known in that area for 70 years, having been described very early in the twentieth century in flounders taken from the Irish Sea. Another independent study of lymphocystis in North Sea plaice also concluded that pollution was not a likely cause of high prevalences.[3]

A lymphocystis epizootic with over 50% prevalence was also reported from flatfish in the North Sea at about the same time (1970),[4] and earlier epizootics have occurred in European waters. A lymphocystis epizootic also occurred in the late 1960s in American plaice, *Hippoglossoides platessoides,* from the Grand Banks of Newfoundland.[5] Several possible explanations were offered for the outbreak, including the possibility that the disease is constantly present in the population and may increase in intensity periodically. Earlier in the century, annual lymphocystis prevalences of 11% were reported in flounders from the Murmansk coast, and infections as high as 12% were noted in the same species from the Öresund.[6] None of these outbreaks seems to have any apparent association with environmental contamination.

Despite inconclusive attempts to relate lymphocystis epizootics in flatfish to specific environmental factors, including pollutants, there are some observations of the disease in fish from the Gulf of Mexico that reopen the issue. Lymphocystis was seen in a 1970 survey in Atlantic croaker and sand sea trout, *Cynoscion arenarius,* from the Mississippi coast of the Gulf of Mexico, and the observation was made that "the pollution load was much greater in estuarine systems where lymphocystis was encountered".[7] Marked increases in lymphocystis prevalences were seen in a resurvey in 1976 in Atlantic croakers from the Mississippi coast, with as high as 50% infected fish in some trawl catches. The investigators concluded that "prevalence appears to relate to rainfall, suggesting that toxicants, salinity, or enriched water could play a major role in infections".[8]

Lymphocystis in striped bass, *Morone saxatilis,* on the east coast of the United States seems to have some tenuous association with heated effluents. Unpublished surveys by the staff of the Sandy Hook (NJ) Laboratory of the National Marine Fisheries Service found high prevalences of lymphocystis in limited samples of striped bass overwintering in the heated effluent of a Long Island electric generating station (Northport, NY). The disease is considered rare in striped bass, and its unusual abundance in a localized population may well be related to the abnormally high winter temperatures in which the population existed or to abnormal crowding, with a consequent increase in stress and ease of transfer of the pathogen. The high temperatures may promote survival or transfer of the pathogen or lower resistance of the host, or they may provoke latent infections into recognizable disease. Lymphocystis is considered to be highly infectious; initial lesions often develop where injuries to the fish have occurred; and lymphocystis virus reaches peak infectivity when water temperatures are high. Some or all of these factors may be important in causing the high prevalences observed in overwintering populations of striped bass. An important concern about diseases such as lymphocystis in fish populations overwintering in heated effluents is that a focus of infection may be provided for incoming spring migrants.

Facultative Pathogens: Ulcerations Caused by Species of the Bacterial Genus Vibrio and Other Microorganisms

Vibrio and similar bacterial infections have been implicated in a number of reports of ulcerations in fish. In fact, ulcerations with bacterial etiology are among the commonest abnormalities in fish from polluted waters. Ulcers may be superficial or penetrating. Where bacterial isolations have been made from ulcerated tissue, *Vibrio anguillarum* has been by far the most predominant organism, with other bacterial groups in lesser abundance. It seems to be a reasonable generalization that many of the infections that produce grossly visible ulcerations in fish are bacterial (although viruses and fungi have been implicated in a few instances), and are often due to pathogens of the genera *Vibrio, Pseudomonas,* or *Aeromonas*. Ulceration often begins with scale loss or formation of small papules, followed by sloughing of the skin, exposing the underlying muscles, which may also be destroyed (Figure 14). Bacterial ulcers may have rough or raised irregular margins and will often be hemorrhagic. Ulcers may or may not be associated with fin erosion.

Epizootic ulcerative syndromes have been reported with increasing frequency in fish from many parts of the world, including the east coast of the United States.[9] Primary causes are uncertain for many outbreaks, although viruses, bacteria, fungi, and other pathogens have been proposed in specific geographic locations. Environmental stress, often as a consequence of pollution, has also been implicated in at least some of the reported epizootics.

The ulcerative lesions do not constitute a single disease entity, since their characteristics (and causes) may be quite different in different host species and areas. Such lesions can be considered as generalized responses of fish to infection and/or abnormal environmental conditions. Types of ulcerations have been described, some with several developmental stages, and mortalities have been observed in some outbreaks.

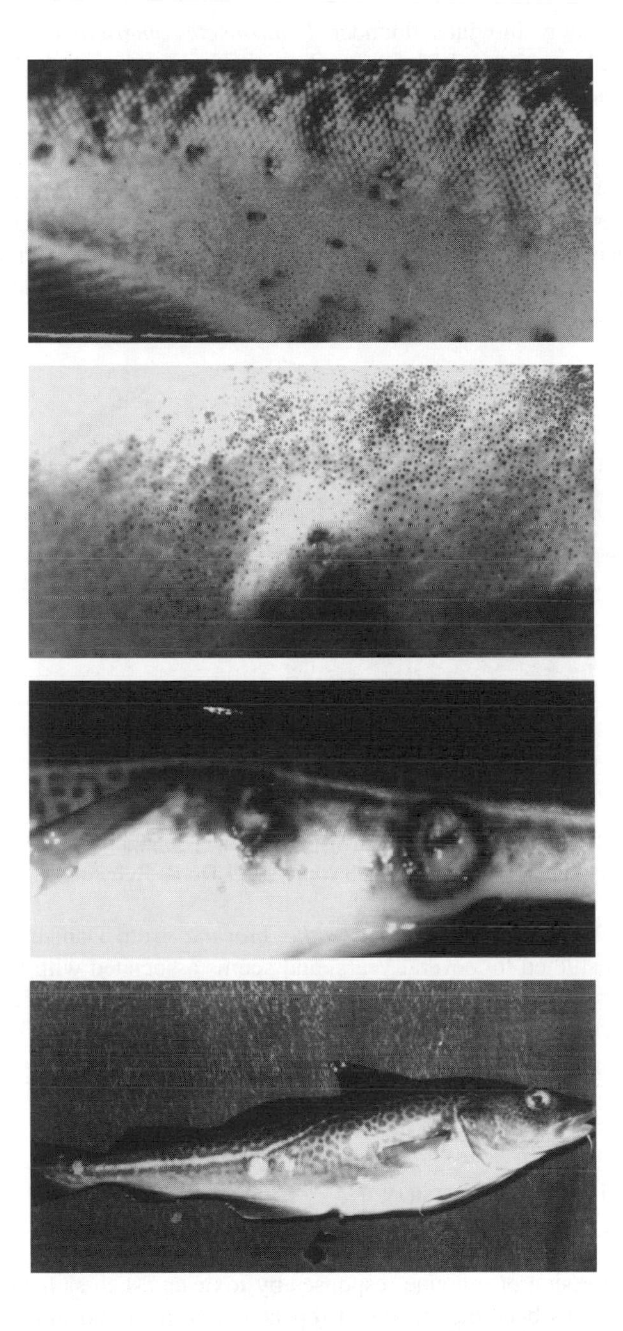

FIGURE 14. Ulcerations in Danish cod: (1) papulovesicular stage, (2) early ulcerative stage, (3) late ulcerative stage, (4) healing stage. (From Christensen, 1980,[12] through the courtesy of Dr. I. Dalsgaard, Copyright ICES. With permission.)

Ulcerations in winter flounder, *Pleuronectes americanus,* in Narragansett Bay have been reported. The acute ulcerative lesions were thought to be caused by the bacterial fish pathogen *Vibrio anguillarum,* and ulcers were induced in fish exposed experimentally to cultured *V. anguillarum* isolates.[10] A subsequent report also described systemic bacterial infections and ulcerations of the tail and dorsal muscles of summer flounder, *Paralichthys dentatus,* from Connecticut waters.[11] A highly pathogenic *Vibrio* species was isolated, and experimental infections were produced by subcutaneous inoculation and by seeding holding tanks with the bacteria. Ulcers and subcutaneous hemorrhages along the bases of fins characterized the infections (Figure 15).

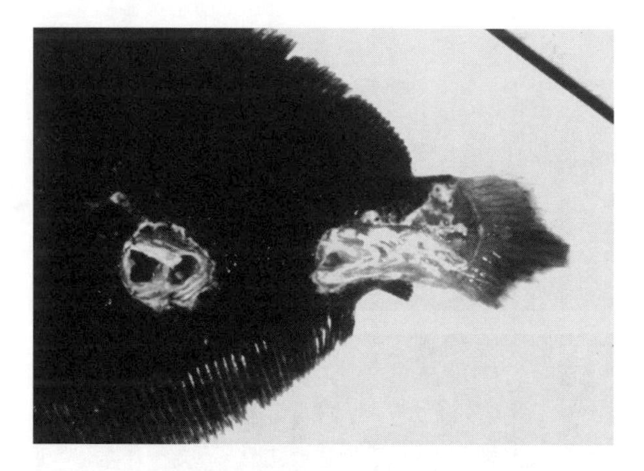

FIGURE 15. Example of ulcerations in fish. Lesion in summer flounder resulting from spontaneous *Vibrio* infections. (Photograph courtesy of Dr. R. Robohm.)

An ulcer syndrome in cod, *Gadus morhua,* from Danish coastal waters has been studied for several years, and seems associated with localized areas of severe pollution.[12] *Vibrio anguillarum* and an *Aeromonas* species have been implicated, but, additionally, several viruses have also been isolated from the lesions, and the suggestion was made that the bacteria may be secondary invaders.

The Role of Immunosuppression in Outbreaks of Infectious Diseases

Suppression of immune responses by toxicants such as heavy metals and pesticides has been demonstrated repeatedly in man and other mammals. It might be expected, therefore, that environmental pollutants could also influence the ability of fish to resist infection by reducing the effectiveness of external and internal defense mechanisms. There is some evidence that this is so. Changes in the principal external defenses — the mucus secretions of fish have been observed, and some information is available about contaminant influences on internal defenses, principally through suppression of immune responses. Stress from contaminants can affect internal resistance to infection in fish by causing a decrease in phagocytic activity of leucocytes or a decrease in antibody synthesis. Both mechanisms have been demonstrated experimentally.

One of the best pieces of information about suppression of host responses was derived from a multidisciplinary experimental study of the effects of short-term sublethal exposure to cadmium on a fish, the cunner, *Tautogolabrus adspersus*.[13] The study included chemical analyses of tissue uptake, physiological and biochemical effects, histopathological changes in tissues, and effects on the immune system. The investigators who were responsible for the immunology found that exposure to 12 ppm cadmium affected the cellular (phagocyte) response, but not the humoral (body fluid) response. The rate of bacterial ingestion by phagocytes of the liver and spleen was increased, but the rate of bacterial destruction within the phagocytes was decreased significantly. No change was observed in the antibody response of immunized control and experimental fish, as determined by hemagglutination (blood cell clumping) techniques. The investigators postulated that cadmium may prevent delivery of lysosomal (cellular enzyme) substances to the phagocytic vacuole or may inhibit the action of these substances on bacteria, but that cadmium does not seem to inhibit antibody synthesis by lymphocytic blood cells. Cadmium and possibly other pollutants may thus affect fish populations by causing phagocytic dysfunction, reducing the competence of fish to destroy bacterial and possibly and other pathogens.

The effects of sublethal exposure to copper on the immune responses of juvenile coho salmon, *Oncorhynchus kisutch,* were examined in a related study.[14] At copper levels of 18 μg/l, serum antibody titers in fingerlings injected intraperitoneally with *Vibrio anguillarum* bacteria were significantly lower than those of controls.

Reduction in immunological competence may well have been involved in observed outbreaks of vibriosis *(V. anguillarum)* in eels exposed to copper, and in epizootics of *Aeromonas liquefaciens* in Atlantic salmon, *Salmo salar,* exposed to copper and zinc contamination, although in neither instance were antibody titers determined.[15] In the latter study, *A. liquefaciens* was described as an opportunistic bacterium that causes disease and mortalities only in fish with lowered resistance.

Immunosuppression by certain pollutants has thus been demonstrated in fish, but the story is not simple. Some recent studies indicate that, for species such as the summer flounder, *Paralichthys dentatus,* antibodies against bacteria common in the environment were higher and existed for a greater diversity of bacteria in fish from polluted waters than in those from an unpolluted reference site — the New York Bight apex versus Great Bay in southern New Jersey.[16] Additionally, selection of high antibody responders among flatfish from the highly polluted New York Bight apex was reported. When tested experimentally against an array of unrelated bacteria, fish taken from polluted waters were found to have higher antibody responses than those from unpolluted waters. This seems to be an example of compensatory population response, in which any reduction in internal defenses (immunosuppression in particular) in individual fish from degraded habitats could be offset by the development of a higher overall population level of immune competence as a result of selection.

POLLUTION AND NONINFECTIOUS DISEASES

Noninfectious diseases include such conditions as environmentally induced skeletal anomalies, genetic abnormalities, physiological malfunctions, metabolic disorders, and some forms of cancer. For purposes of this discussion, examples of noninfectious diseases to be considered include skeletal anomalies, neoplasms (tumors), and neurosensory pathologies.

SKELETAL ANOMALIES

Skeletal deformities, particularly those of the spinal column, are often observed in fish and are the subject of an extensive literature. Such anomalies may be genetic, resulting from mutations or recombinations; epigenetic, acquired during embryonic development; or postembryonic, acquired during larval or postlarval development. Evidence for a hereditary basis exists for some skeletal anomalies, but other evidence points to the effects of environmental factors such as temperature, salinity, dissolved oxygen, radiation, dietary deficiencies, and toxic chemicals.

Increased occurrences of skeletal deformities and anomalies, considered to be pollution-associated, have been reported in a number of fish species from southern California, the British Isles, and Japan. In studies carried out in California, skeletal deformities occurred with greater frequency in samples from areas with significant pollutant stress, and experimental exposure of fry to very low concentrations of DDT (<1 ppb) produced anomalies in fin rays.[17]

The studies conducted in California two decades ago provide what is probably the most convincing observational evidence for environmental influences on induction of skeletal abnormalities in marine fish. Significantly higher prevalences of anomalies, particularly in the gill support structure, occurred in samples of barred sand bass, *Paralabrax nebulifer,* from the southern California Bight (off Los Angeles and San Diego) than from the less-polluted Baja California coast. The anomalies increased in frequency and severity with increasing size of the fish, and an association with disturbed calcium metabolism was suggested. The investigator pointed to the high chlorinated hydrocarbon and heavy metal levels that characterized the California coastal area, but emphasized that a causal relationship with increased prevalence of anomalies had not been established. However, the suggestion of a possible causal relationship between high environmental levels of chlorinated hydrocarbons and heavy metals (both of which are known to interfere with calcium metabolism), and skeletal anomalies in fish seems reasonable in view of experimental evidence from a wide range of vertebrates.

As an extension of the California study of barred sand bass, additional observations were made on two other Pacific coastal species — California grunion, *Leuresthes tenuis,* and barred surfperch, *Amphistichus argenteus* — in which gill support structural anomalies increased in frequency with age, and were "virtually restricted to [samples from] fishes from southern California." This finding in three different species reduces the likelihood that frequency differences could be attributable to inherited subpopulation differences in one of the three species studied. While the deformed gills were the most prevalent anomalies observed in southern California barred sand bass, other abnormalities (pugheadedness, cranial asymmetries, deformed vertebrae, and fin anomalies) occurred and were associated directly in frequency and severity with the gill deformity.

Several reports from Japan refer to high and increasing occurrences of skeletal anomalies in fish.[18] Increasing numbers of malformed sweetfish or ayu, *Plecoglossus altivelis,* have been seen in rivers and culture farms. Skeletal abnormalities in mullet and eight other species from the Inland Sea of Japan have been reported. In a recent study, increasing prevalences of two skeletal syndromes (bent spines and fused vertebrae) were seen as the investigator (T. Matsusato) ascended an estuary (Kurose River) contaminated upstream with chlorinated hydrocarbons and organophosphates.[19] The sample consisted of 28,000 fish of 68 species. The same investigator also summarized all reports of skeletal anomalies — especially spinal fractures — in wild fish of Japan, concluding that occurrences were nationwide, with the highest prevalences in agricultural rather than industrial areas, possibly because of pesticide contamination of habitats.

Deformed fin rays and associated skeletal abnormalities have been observed repeatedly in winter flounder from the highly polluted waters of the New York Bight.[20] Related anomalies, in the form of disruption in normal scale patterns and even scale *reversal,* have been noted in samples from polluted waters of Biscayne Bay, Florida. The presence and frequency of such scale pattern anomalies, along with skeletal abnormalities, may be good indicators of the extent of environmental degradation.

Experimental evidence exists for induction of skeletal abnormalities by exposure of early life history stages to environmental contaminants. Severe scoliosis (a deformity of the vertebral column) and associated pathology were induced in the sheepshead minnow, *Cyprinodon variegatus,* exposed to the organochloride pesticide Kepone.[21] The investigators concluded that scoliosis was a secondary effect of Kepone toxicity, with the nervous system or calcium metabolism the primary target. In related studies, the herbicide Trifluralin (Treflan) induced extensive proliferation of bone-forming cells in vertebrae of sheepshead minnows when life history stages from fertilized egg to 28-day juveniles were exposed to 25 to 50 ppb.[22] Centra of vertebrae, thickened by active bone-forming cells, increased in size up to 10 to 30 times their normal dimensions — a striking sublethal effect.

Neoplasms (Tumors)

Circumstantial evidence associating environmental contamination with neoplasms (tumors) in fish has accumulated from a number of studies, beginning more than half a century ago:

1. Epitheliomas of lips and mouth were found in 1941 in 166 catfish, *Ameiurus nebulosus,* taken from the Delaware and Schuylkill Rivers near Philadelphia.[23] The rivers were grossly polluted. Tumors of this type may result from mechanical, infectious, or chemical irritation. Catfish from other less-polluted areas did not have a high prevalence of tumors. The investigators considered the possibility that the lesions were induced by chemical carcinogens in the water. The lesions developed into carcinomas, some of which were invasive.
2. Papillomas of lips and mouth were found in 10 of 353 white croakers, *Genyonemus lineatus,* from Santa Monica Bay, California, in 1957.[24] Fish were taken 2 m from an ocean outfall. No tumors were found in 1,116 croakers from unspecified nonpolluted waters 70 km away.
3. Cauliflower disease (stomatopapilloma) (see Figure 1) has been increasing in prevalence in eels, *Anguilla anguilla,* from the Baltic Sea and adjacent waters since 1957.[25] The pattern of spread and the high prevalences indicate an infectious process (viral arrays have been seen) or progressive accumulation of industrial contaminants such as fuel oil and smelter wastes (known to contain carcinogenic hydrocarbons such as benzo[*a*]pyrene and heavy metals such as arsenic).
4. Small (10–15 cm) Dover sole, *Microstomus pacificus,* from Santa Monica Bay were seen with tumors in 1964.[26] Fish above 15 cm did not have tumors. Additionally, white croakers from Santa Monica and Los Angeles–Long Beach were found with papillomas of the lips, and papillomas were observed on tongue soles, cusk eels, and Pacific sand dabs. Such tumors were not seen on fish from unpolluted areas. Dover sole with epidermal papillomas have since been collected off Baja California as far south as Cedros Island.[27] The prevalence of lip tumors in white croakers from Santa Monica and the Palos Verdes shelf has been <1% since 1970.
5. Of nearly 16,000 English sole sampled from San Francisco Bay in 1969, 12% had epidermal papillomas, with as many as 33 tumors per fish.[28] Prevalences of tumorous fish in the northern part of the bay were twice those in the southern part. The greatest concentration of industrial waste discharges, especially petrochemicals, existed in the northern part of the bay. A later survey (in 1971), however, failed to confirm the areal difference in tumor abundance.[29]
6. Wartlike tumors histologically resembling fibromas were seen in mullet, *Mugil cephalus,* from Biscayne Bay (FL) in 1969–1970[30] (Figure 16). Other fibrous tumors have been reported since then in mullet from the Gulf of Mexico.

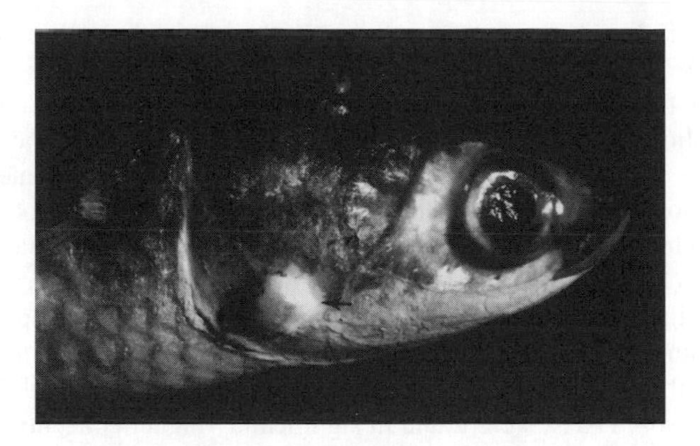

FIGURE 16. Fibrous tumor in mullet.

Chemical examinations of fish tissues from grossly contaminated habitats frequently disclose high levels of heavy metals, chlorinated hydrocarbons, and polycyclic aromatic hydrocarbons (PAHs) in critical metabolic sites such as livers. Pathological changes have been seen in cells and tissues of those organs; such changes may, in fish from severely degraded areas, be neoplastic (cancerous). Probably the best evidence for a relationship of tumors and coastal/estuarine pollution can be found in several studies of hepatomas (liver tumors) in fish.

Hepatomas in Atlantic hagfish, *Myxine glutinosa,* were studied in the early 1970s.[31] Hagfish from a polluted Swedish estuary had prevalences of neoplastic livers that were five times higher than those sampled in the open sea, and a possible association with polychlorinated biphenyl (PCB) contamination in the estuary was suggested. Soon after that, 25% of the livers of Atlantic tomcod, *Microgadus tomcod,* from the polluted Hudson River estuary were found to contain neoplastic nodules and hepatomas (hepatocellular carcinomas), with the highest prevalences in older fish.[32] The investigators suggested a possible association of hepatomas with elevated PCB levels in the Hudson River and in the liver tissues of some specimens. (A reexamination 20 years later disclosed even higher prevalences of the lesions.)

Then, in the late 1970s and the 1980s, the association between progressively severe liver pathology and several types of liver neoplasms with badly degraded coastal/estuarine waters became more evident. Several reports demonstrated this relationship. In the North Atlantic, winter flounder, *Pleuronectes americanus,* from severely degraded areas on the east coast of the United States (New Haven Harbor, upper Narragansett Bay, Boston Harbor) were reported to have prevalences of 3.4 to 7.5% tumors (classified technically as hepatocarcinomas or cholangiocarcinomas[33] (Figure 17). In the North Pacific, prevalences of liver tumors as high as 16% were found in English sole, *Parophrys vetulus,* from the polluted Duwamish River near Seattle, Washington — a river known to contain high levels of PCBs and many other hydrocarbons. In subsequent studies, liver neoplasms were also found (in lesser numbers) in rock sole, *Lepidopsetta bilineata,* starry flounder, *Platichthys stellatus,* and Pacific staghorn sculpin, *Leptocottus armatus*.[34] The neoplasms were found in samples from several polluted sites in the Puget Sound area (Commencement Bay, Eliot Bay, Everett Harbor, and Mukilteo Harbor). The latter study included a detailed analysis of contaminants in tissues and sediments. Positive correlations were obtained between neoplasm prevalence in bottom-dwelling fish and levels of "certain individual groups of sediment-associated chemicals" (PAHs, chlorinated hydrocarbons, and heavy metals). As part of the same study, significant positive correlations were found between prevalences of neoplasms and other lesions and the concentrations of metabolites of aromatic hydrocarbons in bile of the fish.[35] Additionally, organic free radicals, possibly derived from PAHs, were found in significantly higher concentrations in liver cells of fish with lesions than in normal individuals.[36] A clear cause-and-effect relationship was not claimed by the authors of these studies in the Pacific Northwest, who stated that limitations included present inability to identify all sediment-associated contaminants and uncertainty about the synergistic/antagonistic interactions among classes of chemicals.

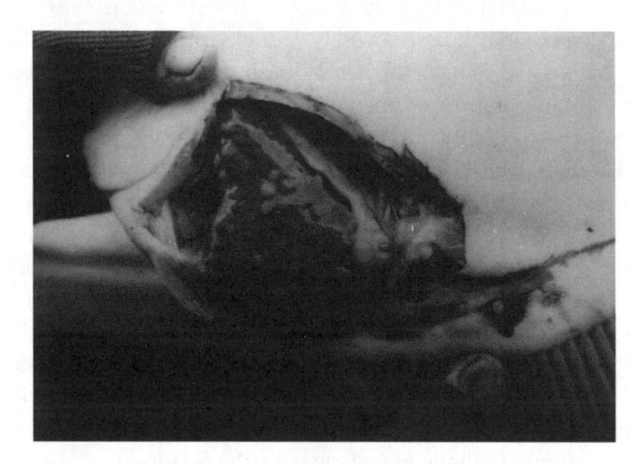

FIGURE 17. Gross lesions in liver of winter flounder, *Pleuronectes americanus*. (Photograph courtesy of Dr. R. A. Murchelano.)

Liver lesions described as tumors have also been reported recently from flounder, *Platichthys flesus,* and dab, *Limanda limanda,* from Dutch coastal waters.[37] Prevalences of gross lesions in fish older than 3 or 4 years were locally as high as 40% and were higher in samples from polluted areas than those from less polluted waters. In other studies, livers of ruffe, *Gymnocephalus cernua,* from the Elbe estuary in Germany were found to have various lesions, including neoplastic nodules. Prevalences in larger sexually mature fish were as high as 32%.[38]

What seems to be emerging from a number of studies of different fish species in different parts of the world is a sequence of histopathological changes in livers, including preneoplastic changes in liver cells. As an example, four kinds of liver lesions have been recognized in studies of flatfish tumors in the Pacific Northwest.[39] These lesions appear to be related sequentially to the genesis of tumors in English sole. The progression of pathological changes in livers seems roughly correlated with the extent of estuarine degradation and the length of residence of fish in the estuary.

In addition to field observations and subsequent correlations of liver tumors and pollution, there is some experimental evidence for the induction of liver neoplasms in marine/euryhaline fish by known carcinogens of higher vertebrates. The Japanese medaka, *Oryzias latipes,* a fish capable of surviving in a wide range of salinities, and sensitive to carcinogens, has been used in several studies. Liver, eye, and other tumors developed after brief exposure of early life history stages.[40] Medaka juveniles exposed to benzo[a]pyrene (BaP) developed what appeared to be sequential lesions terminating in carcinomas. Another species, the sheepshead minnow, *Cyprinodon variegatus,* also developed liver and other neoplasms after exposure to carcinogens.

Neurosensory pathologies

Neurosensory systems of fish, especially chemoreceptors, can be affected by exposure to polluted environments. Vulnerability to chemical toxicants has been demonstrated experimentally in olfactory (smell chemoreceptors) and lateral line (mechanoreceptors) cells of marine species. Studies using minnows — the killifish, *Fundulus heteroclitus,* and the silverside, *Menidia menidia* — exposed to copper, silver, and mercury disclosed severe neural pathology in two forms:

- Degeneration of all cellular elements of the anterior lateral line system
- Severe degenerative changes in olfactory organs, especially necrosis of olfactory epithelium

In other studies, the olfactory responses of two species of salmon were found to be inhibited by experimental exposure to copper and mercury.

Effects of neurotoxic chemicals can be potentially lethal in polluted habitats. Loss of olfactory sense could interfere with spawning migrations of anadromous species. Feeding behavior may be severely disrupted by impairment of lateral line and olfactory sensory capabilities. Additionally, toxic effects of pollutants on sensory organ systems, in the words of one investigator, "... are significant even if they do not cause permanent neurological damage, for [even] a temporary disability that prevents an organism from relating to a viable environment for only moments can be disastrous."[41] This cogent observation can be extended to effects on larval survival. In one example, larval winter flounder, *Pleuronectes americanus,* and haddock, *Melanogrammus aeglefinus,* exposed to copper developed moderate to severe olfactory lesions that could interfere with feeding and other behavior and, if prolonged, could be fatal.

Petroleum hydrocarbons have also been shown to damage the olfactory epithelium of fish. Variable pathological changes were observed in the silverside, in which exposure to crude oil and its fractions caused cell proliferation and dilation and congestion of blood vessels. In later studies, degenerative changes occurred in the chemosensory ciliary hairs of the olfactory organs of larval sand sole, *Psettichthys melanostictus,* exposed to the water-soluble fraction of crude oil.[42] According to the investigator, "the degree of structural alteration observed indicated severe damage to the receptor organelles."

Damage to fish eye lenses can also result from exposure to petroleum hydrocarbons. Long-term (6-month) exposure of cunner, *Tautogolabrus adspersus,* to an oil slick and a long-term (8-month) diet contaminated with crude oil resulted in grossly observable lens opacity in rainbow trout, *Oncorhynchus mykiss*.[43] Changes in fluid content of lens fiber cells of trout, thought to indicate early stages in cataract formation, were also reported in fish that were exposed to petroleum.

In field studies of other species, cataracts were observed in fish from heavily contaminated areas of the Elizabeth River in Virginia.[44] Prevalences in the most contaminated zones were 21% in weakfish, *Cynoscion regalis,* 18% in croaker, *Micropogonias undulatus,* and 10% in spot, *Leiostomus xanthurus*. The eye abnormalities were reported to increase in frequency at stations whose sediments contained high levels of PAHs, and cataracts were induced by experimental exposure of spot to PAH-contaminated Elizabeth River sediments.[45]

Damage to the brain and retina of California surf smelt, *Hypomesus pretiosus,* embryos resulted from experimental exposure to crude oil fractions.[46] Necrotic neurons were seen in the developing forebrain and the neuronal layer of the retina of 27-day embryos after repeated exposures. Damage to retinal receptor cells took the form of cytoplasmic vacuolation and lysis of mitochondria. Mitochondrial disruption in brain cells of larval Pacific herring, *Clupea harengus pallasi,* also resulted from exposure to crude oil.

Neural tissues of some fish seem to be particularly affected by certain petroleum hydrocarbons, and fish brain tissue can bioaccumulate relatively high levels. Benzene, naphthalene, methylnaphthalene, and anthracene were found in several studies to have been sequestered in the brains of rainbow trout, *Oncorhynchus mykiss,* and coho salmon, *Oncorhynchus kisutch*.[47]

POLLUTION AND DISEASES OF "MIXED" OR UNCERTAIN CAUSATION

Apart from those disease conditions that are clearly of infectious or noninfectious etiology, there are others that may be the result of either one or the other process, or of a synergism between the two processes. Included in this category would be skin lesions, especially a condition known as fin erosion. In many instances, it is probably the combination of an infectious agent and environmental stressor that eventually produces disease and in some instances mortality.

Some of the clearest and statistically most defensible associations between pollution and disease are those signaled by the presence of skin lesions in which microbial pathogens are often, but not always, implicated. Probably the best known but least understood disease of fish from polluted waters is a nonspecific condition known as "fin rot" or "fin erosion," a syndrome that seems clearly associated with degraded estuarine or coastal environments — to the extent that it has been proposed as an index of pollutant-induced disease. Fin erosion has been reported from the New York Bight, the California coast, Biscayne Bay and Escambia Bay in Florida, the western Gulf of Mexico, the Irish Sea, and the coast of Japan.

Fin erosion seems to occur in at least two types: one in bottom-dwelling fish, where damage to fins seems site-specific and probably related to direct contact with contaminated sediments (Figure 18), and another in free-swimming nearshore species, characterized by more generalized fin erosion, but with predominant involvement of the tail fin.

FIGURE 18. Fin erosion in summer flounder.

Surveys along the Middle Atlantic coast in the 1970s and 1980s disclosed high prevalences (up to 38%) of fin rot in samples of trawled marine fish — especially from the New York Bight. Investigators found 22 affected species. While bacteria of the genera *Vibrio, Aeromonas,* and *Pseudomonas* were isolated frequently from abnormal fish, a definite bacterial cause has not been established. Fin rot disease was significantly more abundant in the New York Bight apex (the coastal area from Montauk, New York to Cape May, New Jersey) — an area of greatest environmental damage — than in any comparable coastal area from Block Island, Rhode Island, to Cape Hatteras, North Carolina.[48] A correlation between high fin rot prevalence and high coliform counts in sediments has emerged, as has a correlation between high fin rot prevalences and high heavy metal levels in sediments.[49] The disease has been produced experimentally by exposure of fish to polluted sediments.

Fin erosion, with its associated mortalities, was reported several decades ago in Atlantic croaker, *Micropogon undulatus,* and spot, *Leiostomus xanthurus,* from Escambia Bay, Florida.[50] The disease syndrome and mortalities were observed for several years during periods of high temperature and low dissolved oxygen. Escambia Bay had been polluted by PCBs for a number of years.

Information from research in southern California also indicates an association of fin erosion with degraded habitats; relevant statements from investigators are: "The incidence of fin erosion was high in areas with high concentrations of wastewater constituents in the sediments...." "Although there is a definite association between fin erosion and wastewater discharges, the causal factors are unknown." "Nearly half of the 72 species caught off the Palos Verdes Peninsula were affected with this syndrome [eroded fins]".[51]

Some species are either more resistant to fin erosion or are exposed differentially to toxic substances in water or sediments. A study in a heavily polluted arm of Puget Sound (the Duwamish River) in which over 6,000 fish of 29 species were examined, disclosed fin erosion only in starry flounder, *Platichthys stellatus,* and English sole, *Parophrys vetulus.*[52] Average prevalences were 8 and 0.5%, respectively. The investigators also briefly described observations of liver pathology in starry flounder from the area where fin erosion was common. Subsequent studies disclosed that all starry flounders from the Duwamish estuary with fin erosion also had severe liver lesions, and a correlation of liver damage and fin erosion was found also in Dover sole, *Microstomus pacificus,* from the California coast.[53]

Recent Japanese publications have reported fin erosion in fish from polluted bays.[54] As many as 60% of all stargazers, *Uranoscopus japonicus,* sampled from Suruga Bay had evidence of disintegrating caudal and pectoral fins. Six other species also had abnormal fins.

An increase in occurrence of fin erosion and other epidermal lesions (ulcers and lymphocystis) in flatfish from the Irish Sea has been reported since 1970.[55] Fin damage, unknown before 1970, was observed in plaice, *Pleuronectes platessa,* and dab, *Limanda limanda,* taking the form of erosion or total loss of caudal and lateral fins. The investigators pointed to ocean dumping of toxic wastes, particularly of PCBs, as a possible factor contributing to observed prevalences of epidermal lesions, but no clear relationship was demonstrated.

The possible role of environmental chemical contamination in the etiology of fin erosion has emerged more clearly as additional studies have been reported. Fish from the New York Bight, reported in a long series of studies, exist in a highly contaminated area, with chemicals such as heavy metals and petroleum residues in sediments far above background levels. In California, DDT was found to be significantly higher in fish with fin erosion, and PCB levels were slightly higher in such fish than in normal individuals.[56] Both contaminants were much higher in Palos Verdes (a polluted area) fish than in fish from a distant reference area (Dana Point). Abnormally high concentrations of

PCBs were measured in English sole and starry flounders with fin erosion from the Duwamish River in Washington.

Several authors have hypothesized that fin erosion in flatfish may be initiated by direct contact of tissues with contaminated sediments.[57] Toxic substances (for example, sulfides, heavy metals, chlorinated hydrocarbons) could remove or modify the protective mucus coat and expose epithelial tissues to the chemicals. As an example of this, Dover sole from the California coast with severe fin erosion were found to produce much less mucus than normal fish.

It seems quite likely that the "fin erosion" syndrome in fish includes participation of some or all of these factors: (1) chemical stressors, possibly acting on mucus and/or epithelial cells; (2) marginal dissolved oxygen concentrations, possibly enhanced by a sulfide-rich environment; (3) biochemical changes in cell metabolism; and (4) secondary bacterial invasion in at least some instances. Recent experimental information tends to support this hypothesis.

Fin erosion can be induced by exposure to pollutants. A series of experiments at the Gulf Breeze (FL) Environmental Research Laboratory of the U.S. Environmental Protection Agency, using the spot, *Leiostomus xanthurus*, demonstrated experimental production of fin rot disease following exposure to 3 to 5 μg/l of PCBs. Mortalities of up to 80% were reported.[58]

Mullets, *Mugil cephalus*, exposed to 4.5 ppm crude oil in brackish water ponds (12 ppt), developed fin erosion within 6 to 8 days (Figure 19).[59] Lesions were often hemorrhagic, and a tentative *Vibrio* sp. was isolated consistently from surfaces of affected fish but was rarely found systemically. Fin regeneration occurred in most experimental fish within 2 months after exposure.

Experimental induction of fin erosion followed exposure to several other contaminant chemicals. Chronic exposure of fingerling rainbow trout, *Oncorhynchus mykiss*, to lead caused a variety of grossly visible abnormalities, including fin erosion; and chronic exposure of minnows, *Phoxinus phoxinus*, to zinc and cadmium resulted in similar abnormalities.[60]

It seems likely that generalized disease signs, such as fin erosion (and probably other skin lesions such as ulcerations, papillomas, gill hyperplasia, and lymphocystis), may be characteristic of fish that exist in degraded habitats, where environmental stressors (especially toxic chemicals, low dissolved oxygen, and high microbial populations) are present. The extent and nature of external manifestations probably vary with resistance of the particular species, degree of environmental degradation, and length of exposure time.

CONCLUSIONS

Disease can be viewed and described from several perspectives: as an imbalance in the dynamic interplay of host, pathogen, and environment; as a readily recognizable abnormality resulting from severe or prolonged stress; or as inability of a host to resist successfully invasion by a potential pathogen.

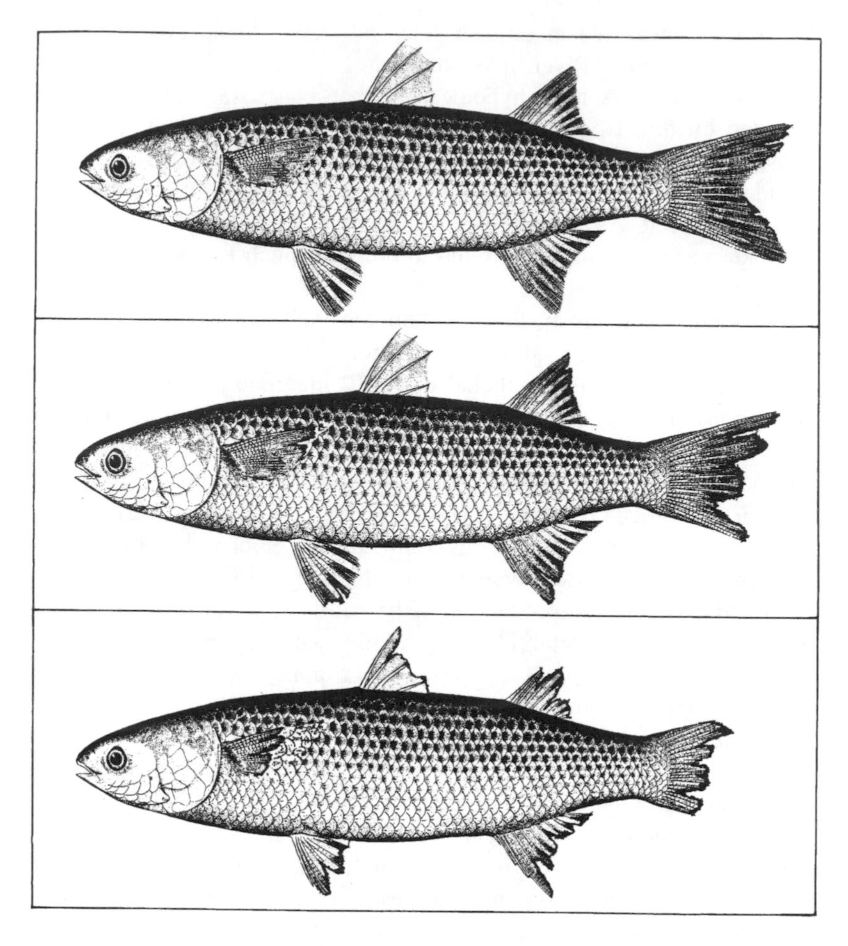

FIGURE 19. Fin erosion produced experimentally in mullet by exposure to oil. (A) Initial fraying of caudal fin, (B) moderate erosion, and (C) extensive erosion. (Modified from Minchew, C.D. and J.D. Yarbrough, *J. Fish Biol.* 10: 319, 1977.)

The integral role of stress in fish diseases has been reaffirmed repeatedly for the past half century, and new information about pollution as a stressor has been accumulating rapidly, especially during the past two decades. Pollutants can act in a variety of ways: by suppressing the immune responses of fish; by interfering with cell enzyme activity, leading to physical abnormalities in developing young; by altering external surfaces, permitting invasion by pathogens; by providing an organic source for population increases of opportunistic pathogens such as vibrios; or by inducing general debilitation, often leading to infections and death (Figure 20).

Based on the published information considered in preparing this chapter, and recognizing the need for greater depth in many of the investigations, it is possible to propose a number of general conclusions:

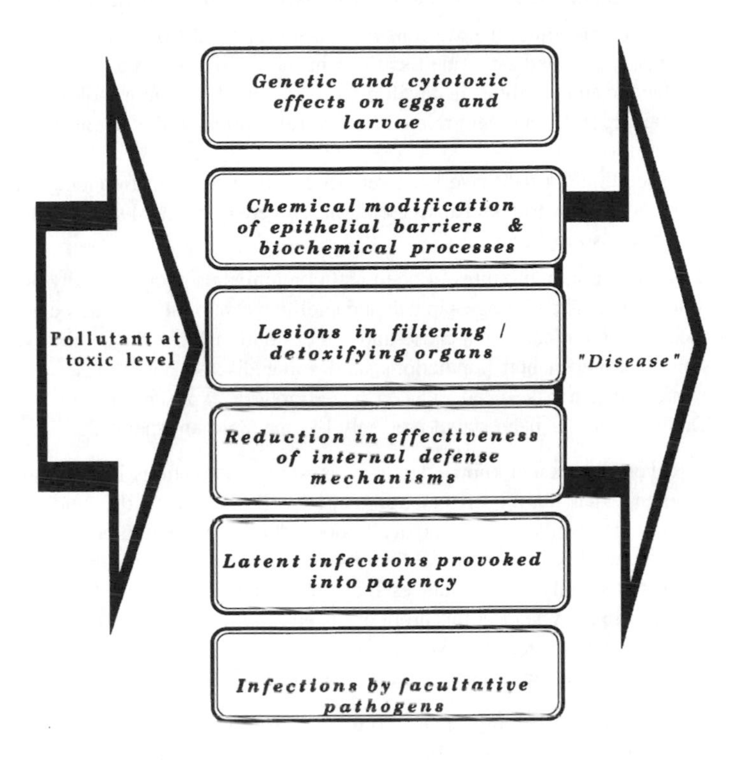

FIGURE 20. Some effects of pollutants on disease processes.

- Disease is a significant limiting factor in coastal/estuarine populations of fish; its effects may be enhanced by stresses resulting from abnormal environmental conditions, including (but not limited to) increases in population abundance of facultative pathogens and toxic chemical concentrations that may affect metabolism and body structures.
- The critical role of environmental stressors, particularly toxic pollutants, in disease is becoming more and more obvious, and the pathways leading to pathology are receiving more attention too. Most of the disease conditions considered in this chapter are now accepted as indicators of stress. Fin erosion, ulcerations, and decreased resistance to facultative pathogens are the best indicators. Such stress-induced abnormalities, which are all consequences of disturbed metabolism, may be augmented by other conditions — such as liver neoplasia — that seem related more directly to the effects of toxic levels of contaminants in the immediate environment, but are of course still indicators of disturbed metabolism.
- Effects of contaminants on living organisms are fundamentally biochemical, resulting in malfunctions in cellular metabolic processes and eventually in

structural and functional changes that become apparent at higher levels of organization and are identified as "disease."

- During the past two decades, important new findings have added significantly to information about pollution and fish diseases. Among them are these:
  1. Tumors and other liver lesions have been reported from flatfish sampled in grossly polluted estuarine locations in the United States and in Europe.
  2. Genetic abnormalities in developing fish embryos and morphological abnormalities in larvae seem related, in several studies, to the extent of chemical pollution.
  3. Ulcerations in fish have been reported with increasing frequency in many parts of the world, and an association with environmental stressors has been hypothesized frequently.
- Because complex mixtures of chemical contaminants occur in badly degraded waters, specific pathologies in fish and shellfish cannot often be associated with specific contaminants in a cause-and-effect relationship. This is, however, feasible in experimental populations, and some disease conditions such as fin erosion in fish have been induced by laboratory exposures to contaminants. The problem has been stated precisely by one German scientist[61]:

  > Given the present knowledge of ecosystems interactions it is unrealistic to expect that marine or aquatic science will be able in the near future to produce results that unequivocally document a connection between specific dysfunctions and specific pollutants. Such causal relationships can be demonstrated only for substances that are tested under controlled laboratory conditions. Whenever this proof has been required or expected for a complex ecosystem, the results have always been debated in the scientific community and therefore have been of no use in a pollution management context.

- Recognizing the validity of this comment, it still seems feasible to establish criteria that *lead toward* establishing cause-and-effect relationships. Other German workers[62] have proposed six criteria for this:
  1. A correlation is found between the disease rate and the overall distribution of wastewater within a narrowly delimited area;
  2. Parallel tendencies are detected between disease prevalence and pollution levels over a long time period;
  3. The disease occurs regularly in heavily polluted water but is very seldom, if ever, encountered in waters with little or no pollution;
  4. The same disease signs and rates of infection encountered among wild fish can be produced during long-term experiments employing controlled concentrations of one or more contaminants in the water;
  5. A reduction in the water pollution or in the concentrations of the major contaminants results in a decline in the disease rate, if the condition is reversible;
  6. Those individuals with highest contaminant burdens are those afflicted with the disease.

The investigators then point out the obvious — that satisfaction of any single criterion from this list does not constitute proof of a cause-and-effect relationship, but the *probability* (and *only* the probability) of such a relationship increases as additional criteria are simultaneously fulfilled.

- It should be reemphasized, in the presence of this persistent problem of causality, that extreme caution must be exercised in associating pollutants with disease conditions in fish; a point made in great detail by some United States and European researchers. Many factors other than pollutants may be involved in producing lethal or sublethal effects; among them would be oxygen depletion, abnormal salinities or temperatures, or toxins from dinoflagellate blooms. Additionally, fish species vary greatly in responses to abnormal environmental factors (including anthropogenic chemicals), and specific life stages may be remarkably different in susceptibility to variations in any single factor.
- In examining the available evidence for an association of certain disease conditions and habitat degradation, it is important to keep the quality and quantity of available data in mind. It is easy to err on the side of *extremism* — finding relationships lurking in even moderately disturbed environments; it is equally easy to err on the side of *conservatism* — discounting all observations as coincidental or circumstantial and therefore irrelevant. Examples of suggested associations and correlations of pollution and disease presented in this chapter represent accumulations of data assembled by many observers, but only a few studies (those of genetic anomalies in developing eggs; those of malformed fish larvae; those of fin erosion; and those of liver tumors) have had the necessary broad, multidisciplinary base to be described as "definitive." The point to be stressed here is that some *correlations* of disease prevalence and pollution *have* been made, and the body of data is increasing, but *direct, specific cause-and-effect relationships have not been established,* except in experimental studies.

Despite increasing public concern about the effects of pollution on inhabitants of coastal/estuarine waters, attempts to slow the rate of environmental degradation will undoubtedly be part of a long-term process, since human populations living in coastal areas continue to expand. It is logical to assume, therefore, that pollution-associated disease problems in fish will not disappear; at best, they may be lessened slightly during the coming decades if efforts are made to reduce the cumulative anthropogenic impacts in the most severely polluted areas. Pollution-associated disease conditions have often been recognized and examined in the worst coastal/estuarine environments that humans have done much to create — places like the New York Bight apex, the mouths of the Elbe and Rhine Rivers, the Oslofjord in Norway, the Duwamish River in Puget Sound, Tokyo Harbor, and the Houston ship canal, to mention only a few. Decreases in contamination of these areas should at some point be reflected in reduction in prevalence and severity of many of the disease conditions described in this chapter.

REFERENCES

1. **Perkins, E. J., J. R. S. Gilchrist, and O. J. Abbott.** 1972. Incidence of epidermal lesions in fish of the north-east Irish Sea area, 1971. *Nature* (London) 238: 101–103.
2. **Shelton, R. G. J. and K. W. Wilson.** 1973. On the occurrence of lymphocystis, with notes on other pathological conditions, in the flatfish stocks of the north-east Irish Sea. *Aquaculture* 2: 395–410.
3. **van Banning, P.** 1971. Wratziekte bij platvis. *Visserij* 24: 336–343.
4. **Mann, H.** 1970. Über den Befall der Plattfische der Nordsee mit Lymphocystis. *Ber. Dtsch. Wiss. Komm. Meeresforsch.* 21: 219–223.
5. **Templeman, W.** 1965. Lymphocystis disease in American plaice of the eastern Grand Bank. *J. Fish. Res. Board Can.* 22: 1345–1356.
6. **Awerinzew, S.** 1911. Studien über parasitische Protozoen. V. Einige neue Befunde aus der Entwicklungsgeschichte von *Lymphocystis johnstonei. Woodc. Arch. Protistenkd.* 22: 179–196; **Nordenberg, C.** 1962. Das Vorkommen der Lymphocystiskrankheit bei Scholle und Flunder in Oresund. *K. Fysiogr. Sallsk. Lund Forh.* 43: 17–26.
7. **Christmas, J. Y. and H. D. Howse.** 1970. The occurrence of lymphocystis in *Micropogon undulatus* and *Cynoscion arenarius* from Mississippi estuaries. *Gulf Res. Rep.* 3: 131–154.
8. **Edwards, R. H. and R. M. Overstreet.** 1976. Mesenchymal tumors of some estuarine fishes of the northern Gulf of Mexico. I. Subcutaneous tumors, probably fibrosarcomas, in the striped mullet, *Mugil cephalus. Bull. Mar. Sci.* 26: 33–40.
9. **Sindermann, C. J.** 1988. Epizootic ulcerative syndromes in coastal/estuarine fish. U.S. Department of Commerce, National Marine and Fisheries Service, NOAA Tech. Memo. NMFS-F/NEC-54, 37 pp.
10. **Levin, M. A., R. E. Wolke, and V. J. Cabelli.** 1972. *Vibrio anguillarum* as a cause of disease in winter flounder *(Pseudopleuronectes americanus). Can. J. Microbiol.* 118: 1585–1592.
11. **Robohm, R. A. and C. Brown.** 1977. A new bacterium (presumptive *Vibrio* species) causing ulcers in flatfish. *Abstr. 3, Annu. East. Fish. Health Worksh.*
12. **Jensen, N. J. and J. L. Larsen.** 1976. Forskningsprojet vetrorende marine vibriopopulationers for hold til dratforurening i marine biotoper. *Dan. Vet. Tidsskr.* 58: 521–524; **Jensen, N. J. and J. L. Larsen.** 1977. Sarsygdom hos torsk i danske kystnaere omrader med forskellig vandkvalitet. *Nord. Veterinaermed.* 29 (Suppl. 1): 4; **Christensen, N. O.** 1980. Diseases and anomalies in fish and invertebrates in Danish littoral regions which might be connected with pollution. *Rapp. P.-V. Reun. Cons. Int. Explor. Mer* 179: 103–109.
13. **Calabrese, A., R. S. Collier, and J. E. Miller.** 1974. Physiological response of the cunner, *Tautogolabrus adspersus,* to cadmium. I. Introduction and experimental design. U.S. Department of Commerce, NOAA Tech. Rep. NMFS SSRF-681.
14. **Stevens, D. G.** 1977. Survival and Immune Response of Coho Salmon Exposed to Copper. EPA 600/3–77–031, U.S. Environmental Protection Agency, Washington, D.C.

15. **Rodsaether, M. C., J. Olafsen, J. Raa, K. Myhre, and J. B. Steen.** 1977. Copper as an initiating factor of vibriosis *(Vibrio anguillarum)* in eel *(Anguilla anguilla)*. *J. Fish Biol.* 10: 17–21; **Pippy, J. H. C. and G. M. Hare.** 1969. Relationship of river pollution to bacterial infection in salmon *(Salmo salar)* and suckers *(Catostomus commersoni)*. *Trans. Am. Fish. Soc.* 98: 685–690.
16. **Robohm, R. A., C. Brown, and R. A. Murchelano.** 1979. Comparison of antibodies in marine fish from clean and polluted waters in the New York Bight: relative levels against 36 bacteria. *Appl. Environ. Microbiol.* 38: 248–257; **Robohm, R. A. and D. S. Sparrow.** 1981. Evidence for genetic selection of high antibody responders in summer flounder *(Paralichthys dentatus)* from polluted areas. *Dev. Biol. Stand.* 49: 273–278.
17. **Valentine, D. W., M. E. Soulé, and Y. P. Samollow.** 1973. Asymmetry analysis in fishes: a possible statistical indicator of environmental stress. *Fish. Bull.* 71: 357–370; **Valentine, D. W.** 1975. Skeletal anomalies in marine teleosts, pp. 695–718. in Ribelin, W. E. and G. Migaki (Eds.), *The Pathology of Fishes*. University of Wisconsin Press, Madison.
18. **Komada, N.** 1974. Studies on abnormality of bones in anomalous "Ayu," *Plecoglossus altivelis*. *Gyobo Kenkyu* 8: 127–135 (in Japanese; English summary); **Matsusato, T.** 1973. On the skeletal abnormalities in marine fishes. I. The abnormal marine fishes collected along the coast of Hiroshima Prefecture. *Nansei Kaiku Suisan Kenkyusho Kenkyo Hokoku* 6: 17–58.
19. **Matsusato, T.** 1979. Skeletal abnormalities of marine fishes found in natural waters and their value as environmental indices. *Nansei Kaiku Suisan Kenkyusho Kenkyu Hokoku* 53: 183–252 (in Japanese).
20. **Ziskowski, J. J., V. T. Anderson, and R. A. Murchelano.** 1980. A bent fin ray condition in winter flounder *(Pseudopleuronectes americanus)* from Sandy Hook and Raritan Bays, New Jersey, and Lower Bay, New York. *Copeia* 1980: 895–899.
21. **Couch, J. A., J. T. Winstead, and L. R. Goodman.** 1977. Kepone-induced scoliosis and its histological consequences in fish. *Science* 197: 585–587.
22. **Couch, J. A., J. T. Winstead, D. J. Hansen, and L. R. Goodman.** 1979. Vertebral dysplasia in young fish exposed to the herbicide trifluralin. *J. Fish Dis.* 2: 35–42.
23. **Lucké, B. and H. Schlumberger.** 1941. Transplantable epitheliomas of the lip and mouth of catfish. *J. Exp. Med.* 74: 397–408.
24. **Russell, F. E. and P. Kotin.** 1957. Squamous papilloma in the white croaker. *J. Natl. Cancer Inst.* 18: 857–861.
25. **Koops, H. and H. Mann.** 1969. Die Blumenkohlkrankheit der Aale Vorkommen und Verbreitung der Krankheit. *Arch. Fischereiwiss.* 20: 5–15.
26. **Young, P. H.** 1964. Some effects of sewer effluent on marine life. *Calif. Fish Game* 50: 33–41.
27. **Sherwood, M. J. and A. J. Mearns.** 1976. Occurrence of tumor-bearing Dover sole *(Microstomus pacificus)* off Point Arguello, California and off Baja California, Mexico. *Trans. Am. Fish. Soc.* 105: 561–563; **Mearns, A. J. and M. J. Sherwood.** 1977. Distribution of neoplasms and other diseases in marine fishes relative to the discharge of waste water. *Ann. N.Y. Acad. Sci.* 298: 210–224.
28. **Cooper, R. C. and C. A. Keller.** 1969. Epizootiology of papillomas in English sole, *Parophrys vetulus*. *Natl. Cancer Inst. Monogr.* 31: 173–185.

29. **Kelly, D. L.** 1971. Epidermal papilloma in the English sole, *Parophrys vetulus*. Ph.D. thesis, University of California, Berkeley. 139 pp.
30. **Sindermann C. J.** 1976. Effects of coastal pollution on fish and fisheries — with particular reference to the Middle Atlantic Bight. *Am. Soc. Limnol. Oceanogr.,* Spec. Symp 2: 281–301.
31. **Falkmer, S., S. Marklund, P. E. Mattsson, and C. Rappe.** 1977. Hepatomas and other neoplasms in the Atlantic hagfish *(Myxine glutinosa):* a histopathologic and chemical study. *Ann. N.Y. Acad. Sci.* 298: 342–355.
32. **Smith, C. E., T. H. Peck, R. J. Klauda, and J. B. McLaren.** 1979. Hepatomas in Atlantic tomcod *Microgadus tomcod* (Walbaum) collected in the Hudson River estuary in New York. *J. Fish Dis.* 2: 313–319; **Dey, W. P., T. H. Peck, C. E. Smith, and G.-L. Kreamer.** 1993. Epizootiology of hepatic neoplasia in Atlantic tomcod *(Microgadus tomcod)* from the Hudson River estuary. *Can. J. Fish. Aquat. Sci.* 50: 1897–1907.
33. **Murchelano, R. A. and R. E. Wolke.** 1985. Epizootic carcinoma in the winter flounder, *Pseudopleuronectes americanus*. *Science* 228: 587–589.
34. **Malins, D. C., B. B. McCain, D. W. Brown, S.- L. Chan, M. S. Myers, J. T. Landahl, P. G. Prohaska, A. J. Friedman, L. D. Rhodes, D. G. Burrows, W. D. Gronlund, and H. O. Hodgins.** 1984. Chemical pollutants in sediments and diseases of bottom-dwelling fish in Puget Sound, Washington. *Environ. Sci. Technol.* 18: 705–713.
35. **Krahn, M. M., L. D. Rhodes, M. S. Myers, I. K. Moore, W. D. MacLeod, Jr., and D. C. Malins.** 1986. Associations between metabolites of aromatic compounds in bile and the occurrence of hepatic lesions in English sole *(Parophrys vetulus)* from Puget Sound, Washington. *Arch. Environ. Contam. Toxicol.* 15: 61–67.
36. **Malins, D. C., M. S. Myers, and W. T. Roubal.** 1983. Organic free radicals associated with idiopathic liver lesions of English sole *(Parophrys vetulus)* from polluted marine environments. *Environ. Sci. Technol.* 17: 679–685.
37. **Vethaak, A. D.** 1987. Fish diseases, signals for a diseased environment? pp. 41–61. in Peet, G. (Ed.), *Proceedings of the 2nd North Sea Seminar,* Vol. 2. Werkgroep Nordzee, Rotterdam, Amsterdam.
38. **Peters, N., A. Köhler, and H. Kranz.** 1987. Liver pathology in fishes from the lower Elbe as a consequence of pollution. *Dis. Aquat. Org.* 2: 87–97.
39. **Myers, M. S., L. D. Rhodes, and B. B. McCain.** 1987. Pathologic anatomy and patterns of occurrence of hepatic neoplasms, putative preneoplastic lesions, and other idiopathic conditions in English sole *(Parophrys vetulus)* from Puget Sound, Washington. *J. Natl. Cancer Inst.* 78: 333–363.
40. **Aoki, K. and H. Matsudaira.** 1981. Factors influencing tumorigenesis in the liver after treatment with methylazoxymethanol acetate in a teleost, *Oryzias latipes,* pp. 202–216. in Dawe, C. J., J. C. Harshbarger, S. Konto, T. Sugimura, and S. Takayama (Eds.), *Phyletic Approaches to Cancer.* Japan Scientific Society Press, Tokyo.
41. **Bodammer, J. E.** 1981. The cytopathological effect of copper on the olfactory organs of larval fish (*Pseudopleuronectes americanus* and *Melanogrammus aeglefinus*). *Int. Counc. Explor. Sea,* Doc. C.M.1981/E:46, 11 pp.
42. **Hawkes, J. W.** 1980. The effects of xenobiotics on fish tissues: morphological studies. *Fed. Proc.* 39: 3230–3236.

43. **Hawkes, J. W.** 1977. The effects of petroleum hydrocarbon exposure on the structure of fish tissues, pp. 115–128. in Wolfe, D.A. (Ed.), *Fate and Effects of Petroleum Hydrocarbons in Marine Ecosystems and Organisms*. Pergamon Press, New York; **Payne, J. F., J. W. Kiceniuk, W. R. Squires, and G. L. Fletcher.** 1978. Pathological changes in a marine fish after a 6-month exposure to petroleum. *J. Fish. Res. Board Can.* 35: 665–667.
44. **Bender, M. E., W. J. Hargis, Jr., R. J. Huggett, and M. H. Roberts, Jr.** 1988. Effects of polynuclear aromatic hydrocarbons on fishes and shellfish: an overview of research in Virginia. *Environ. Res.* 24: 237–241.
45. **Hargis, W. J. and D. E. Zwerner.** 1989. Some effects of sediment-borne contaminants on development and cytomorphology of teleost eye-lens epithelial cells and their derivatives. *Mar. Environ. Res.* 28: 399–405.
46. **Hawkes, J. W. and C. M. Stehr.** 1982. Cytopathology of the brain and retina of embryonic surf smelt *(Hypomesus pretiosus)* exposed to crude oil. *Environ. Res.* 27: 164–178.
47. **Roubal, W. T., T. K. Collier, and D. C. Malins.** 1977. Accumulation and metabolism of carbon-14 labelled benzene, naphthalene, and anthracene by young coho salmon *(Oncorhynchus kisutch)*. *Arch. Environ. Contam. Toxicol.* 5: 513–529; **Collier, T. K., M. M. Krahn, and D. C. Malins.** 1980. The disposition of naphthalene and its metabolites in the brain of rainbow trout *(Salmo gairdneri)*. *Environ. Res.* 23: 35–41.
48. **Murchelano, R. A. and J. Ziskowski.** 1976. Fin rot disease studies in the New York Bight. *Am. Soc. Limnol. Oceanogr., Spec. Symp.* 2: 329–336.
49. **Mahoney, J. B., F. H. Midlige, and D. G. Deuel.** 1973. A fin rot disease of marine and euryhaline fishes in the New York Bight. *Trans. Am. Fish. Soc.* 102: 596–605; **Carmody, D. J., J. B. Pearce, and W. E. Yasso.** 1973. Trace metals in sediments of New York Bight. *Mar. Pollut. Bull.* 4: 132–135.
50. **Couch, J. A. and D. R. Nimmo.** 1974. Detection of interactions between natural pathogens and pollutants in aquatic animals, pp. 261–268. in Amborski, R. L., M. A. Hood, and R. R. Miller (Eds.), *Proceedings of the Gulf Coast Regional Symposium on Diseases of Aquatic Animals*. LSU-SG-74–05, Ctr. Wetland Resource, Louisiana State University, Baton Rouge.
51. Southern California Coastal Water Research Project. 1973. The ecology of the southern California Bight: implications for water quality management. Ref. No. SCCWRP TR 104, South. Calif. Coastal Water Res. Proj., El Segundo, CA.
52. **Wellings, S. R., C. E. Alpers, B. B. McCain, and B. S. Miller.** 1976. Fin erosion disease of starry flounder *(Platichthys stellatus)* and English sole *(Parophrys vetulus)* in the estuary of the Duwamish River, Seattle, Washington. *J. Fish. Res. Board Can.* 33: 2577–2586.
53. **Pierce, K. V., B. B. McCain, and M. J. Sherwood.** 1977. Histology of liver tissue from Dover *sole. Annu. Rep. South. Calif. Coastal Water Res. Proj.,* El Segundo, CA, pp. 207–212; **Pierce, K. V., B. B. McCain, and S. R. Wellings.** 1980. Histopathology of abnormal livers and other organs of starry flounder *Platichthys stellatus* (Pallas) from the estuary of the Duwamish River, Seattle, Washington. *J. Fish Dis.* 3: 81–91.
54. **Nakai, Z., M. Kosaka, S. Kudoh, A. Nagai, F. Hayashida, T. Kubota, M. Ogura, T. Mizushima, and I. Uotani.** 1973. Summary report on marine biological studies of Suruga Bay accomplished by Tokai University 1964–72. *J. Fac. Mar. Sci. Technol. Tokai Univ.* 7: 63–117.

55. **Perkins, E. J., J. R. S. Gilchrist, and O. J. Abbott.** 1972. Incidence of epidermal lesions in fish of the north-east Irish Sea area, 1971. *Nature* (Lond.) 238: 101–103.
56. **McDermott, D. J. and M. J. Sherwood.** 1975. Annual report. *Dept. Fish. Mar. Fish Prog., South Calif. Coastal Water Res. Proj.*, El Segundo, CA.
57. **Mearns, A. J. and M. J. Sherwood.** 1974. Environmental aspects of fin erosion and tumors in southern California Dover sole. *Trans. Am. Fish. Soc.* 103: 799–810; **Sherwood, M. J. and A. J. Mearns.** 1977. Environmental significance of fin erosion in southern California demersal fishes. *Ann. N.Y. Acad. Sci.* 298: 177–189.
58. **Couch, J. A.** 1975. Histopathologic effects of pesticides and related chemicals on the livers of fishes, pp. 559–584. in Ribelin, W.E. and G. Migaki (Eds.), *The Pathology of Fishes*. University of Wisconsin Press, Madison.
59. **Minchew, C. D. and J. D. Yarbrough.** 1977. The occurrence of fin rot in mullet *(Mugil cephalus)* associated with crude oil contamination of an estuarine pond-ecosystem. *J. Fish Biol.* 10: 319–323.
60. **Davies, P. H. and J. H. Everhart.** 1973. Effects of chemical variations in aquatic environments. III. Lead toxicity to rainbow trout and testing application factor concept. EPA-R3–73–011C, U.S. Environmental Protection Agency, Washington, D.C. 80 pp; **Bengtsson, B. E.** 1974. Vertebral damage to minnows *Phoxinus phoxinus* exposed to zinc. *Oikos* 25: 134–139; **Bengtsson, B. E.** 1975. Vertebral damage in fish induced by pollutants, pp. 23–30. in Koeman, J.H. and J.J.T.W.A. Strik (Eds.), *Sublethal Effects of Toxic Chemicals on Aquatic Animals*. Elsevier, Amsterdam.
61. **Dethlefsen, V.** 1988. Status report on aquatic pollution problems in Europe. *Aquat. Toxicol.* 11: 259–286.
62. **Peters, N., A. Köhler, and H. Kranz.** 1987. Liver pathology in fishes from the lower Elbe as a consequence of pollution. *Dis. Aquat. Org.* 2: 87–97.

4 Ocean Pollution and Shellfish Diseases

****

BLUE CRABS OF THE PAMLICO ESTUARY

One of the best-tasting of our east coast seafoods — blue crabs (Callinectes sapidus) — occurs in commercial concentrations at many locations along the Atlantic and Gulf coasts of the United States. A center of abundance of this remarkable swimming crab is in the lower reaches of the Pamlico River and Pamlico Sound in North Carolina. Sharing the estuarine area with the crabs, and intruding on an otherwise delightful rural landscape, is a giant chemical industrial complex that sprawls along miles of the shoreline. The company facilities, principally because of their mammoth proportions, serve as a point of interest in an area unmarked by too many tourist attractions (other than the estuary itself), but the industrial processes contribute substantial amounts of toxic contaminants to the estuarine waters — particularly fluoride and cadmium.

During the past decade, much attention has been paid to the highly productive Pamlico estuary, in part because some of the resource species, including fish and crabs, have been found to be abnormal, with external ulcerations in fish and shell disease in crabs. The shell disease, which is considered to be primarily of bacterial origin, is characterized by pitting and erosion of the carapace, often with loss of significant pieces of the exoskeleton. This disease occurs in all crustaceans, but is often associated (in prevalence and severity) with degraded habitats, in areas such as the New York Bight, the Oslofjord in Norway, and elsewhere in the world.

New studies of abnormal blue crabs of the Pamlico estuary suggest a geographic correlation of the occurrence and severity of shell disease with industrial pollution. High levels of pollutants and high prevalences of severe shell disease showed some correlation, indicating (but certainly not proving) a possible cause and effect relationship. The prevailing hypothesis among some scientists is that contaminants from industrial

effluents may interfere with the crab's ability to synthesize chitin and/or to metabolize calcium, thereby disrupting normal shell deposition and maintenance. In some crabs, the amount of shell loss is extensive enough to suggest that the condition can be lethal, but effects on abundance have (as usual) not been determined.

From "Field Notes of a Pollution Watcher"
(C. J. Sindermann, 1991)

****

INTRODUCTION

In the preceding chapter, we considered in great detail the many infectious and noninfectious diseases of marine fish that have at least some tenuous association with coastal/estuarine pollution. Actually, in reexamining that chapter, the evidence for linkages of pollution and fish diseases is substantial, and evidence for cause-and-effect relationships is accumulating constantly.

But now it is time to look at some examples of comparable associations between shellfish (molluscan and crustacean) and certain of their diseases and abnormalities. With these invertebrates we find, as we did with fish, that during the past few decades important findings have added significantly to information about interactions of pollution and disease. Shell disease and associated "black gill disease" of crustaceans have been reported with higher prevalences and with greater severity in polluted habitats. Shell disease is now thought to be a consequence of disturbed chitin synthesis in the presence of overwhelming populations of chitin-destroying microorganisms. With bivalve molluscs, highly toxic chemicals such as tributyltin may interfere with normal calcification of the shell.

The physiological/biochemical pathways through which environmental stressors produce pathological changes in molluscs and crustaceans have not been examined to the same extent as in fish, but analogous pathways are thought to exist. We touched briefly in Chapter 1 on the role of stress in invertebrates, and it seems logical to expand on that discussion here, as a prelude to considering pollution-associated shellfish diseases.

Through a variety of studies, elements of a stress response in molluscs and crustaceans are emerging. Some are general; some relate specifically to molluscs; and some relate specifically to crustaceans. Responses of molluscs include (see Figure 7):

- Slow growth and reduced "scope for growth" (a measure of the potential for body growth and gamete production)
- Decline in physiological condition
- Depression of gametogenic activity

- Destabilization of intracellular lysosomal membranes, resulting in autolysis of cells
- High taurine/glycine (amino acid) ratios in gill and mantle tissues
- Regression and atrophy of the digestive tubule epithelium
- Increase in ceroid (brown cell) aggregations
- Hemocyte infiltration of the tissues
- Mantle recession
- Abnormal shell formation

Stress responses of crustaceans include (see Figure 8):

- Black gill syndrome
- Frequent occurrence of shell disease caused by chitin-destroying microorganisms
- Abnormal muscle opacity
- Abnormal growth of external fouling organisms, especially filamentous bacteria and ciliate protozoans
- Molt retardation
- Presence of gram-positive bacteria in the hemolymph
- Abnormal, disoriented behavior
- Clotting of hemolymph (as a response to the presence of gram-negative endotoxin)

Although most studies of the physiological and morphological consequences of stress have emphasized vertebrates, there are indications that counterpart phenomena may be present in invertebrates as well. For Mollusca, several publications have provided evidence for a generalized stress response syndrome. With the clam, *Mercenaria mercenaria,* for example, a shortened life span, mass mortalities, a reduction in total amino acid and carbohydrate content, and extensive shell invasion by the polychaete worm *Polydora* were consequences of stress.[1] Amino acid ratios, particularly the taurine/glycine ratio in gill and mantle tissues, were considered to be the best indicators of stress. Ratios above 3 indicated stress, with ratios above 5 an indication of acute stress. Exposure of clams to petroleum resulted in black tar–like concretions in amebocytes and in kidney tissues. The concretions collected in the renal sac, plugging the kidney tubules.

Other indices of stress in molluscs have been proposed — particularly a reduced scope for growth and a reduced ratio of oxygen consumed to nitrogen excreted.[2] With mussels, a decline in condition and a depression of gametogenic activity were observed under stress conditions.

Often the outcome of pollution stress is infectious disease, usually with a facultative microbial agent involved. Sometimes, though, the infectious process is not clearly defined, so that the same causative agent may not be isolated consistently, or no infectious agent can be associated with an abnormality in every case. Some examples of infectious and noninfectious diseases are included in the following sections.

POLLUTION AND INFECTIOUS DISEASES

We are learning more all the time about the infectious diseases of shellfish and about the mechanisms utilized by invertebrates to resist infection. We have also achieved considerable understanding of the role that pollution and pollution-induced stress can play in the development of infectious diseases in shellfish. In this part of the chapter, three aspects of infectious diseases will be examined: facultative (opportunistic) pathogens, activated latent viral infections, and reduction in disease resistance resulting from pollution. For each category, examples have been selected from those having the greatest amount of treatment in the scientific literature.

FACULTATIVE PATHOGENS

The preceding chapter on fish diseases emphasized the predominant role of the bacterial genus *Vibrio* in causing disease in stressed individuals. These and many other genera of marine microorganisms are "heterotrophic," in that they may use organic material to obtain energy, or they may attack living tissues as energy sources, thereby acting as pathogens. Our best example of such opportunism, associated with polluted habitats, can be found among the chitin-degrading microorganisms, some of which are exuberant enough to attack living crustaceans, causing shell disease.

SHELL DISEASE IN CRUSTACEANS

A disease condition in Crustacea commonly referred to as "shell disease," "exoskeletal disease," or "shell erosion" has been associated with badly degraded estuarine and coastal waters. This abnormality can be considered in some ways as the invertebrate analogue of fin erosion in fish. Shell disease has been observed in many crustacean species and under many conditions, both natural and artificial[3] (Figure 21). Shell erosion seems to involve activity of chitin-destroying (chitinoclastic) microorganisms, with subsequent secondary infection of underlying tissue by other facultative pathogens. Initial preparation of the exoskeletal substrate by mechanical, chemical, or microbial action probably is significant; thus, high bacterial populations and the presence of contaminant chemicals in polluted environments, as well as extensive fouling of the gills, could combine to make shell disease a common phenomenon and a significant mortality factor in crustaceans inhabiting degraded environments.

Lobsters, *Homarus americanus,* and rock crabs, *Cancer irroratus,* from grossly polluted areas of the New York Bight were found in early studies to be abnormal, with appendage and gill erosion a most common sign.[4] Skeletal erosion occurred principally on the tips of the walking legs, the ventral sides

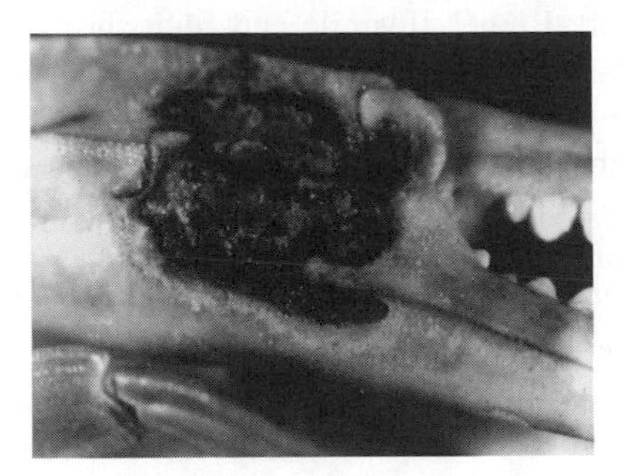

FIGURE 21. Gross signs of shell disease in the claw of a blue crab, *Callinectes sapidus*. (Photograph courtesy of Dr. R.M. Overstreet.)

of chelipeds, exoskeletal spines, gill lamellae, and around areas of exoskeletal articulation where contaminated sediments could accumulate. Gills of crabs and lobsters sampled at the various bight dumpsites were usually clogged with detritus, possessed a dark brown coating, contained localized thickenings, and displayed areas of erosion and necrosis. Similar disease signs were produced experimentally in animals held for six weeks in aquaria containing sediments taken from sewage sludge or dredge spoil disposal sites. Initial discrete areas of exoskeletal erosion became confluent, covering large areas of the exoskeleton, and often parts of diseased appendages were lost. The chitinous covering of the gill filaments was also eroded, and often the underlying tissues became necrotic.

In other early studies,[5] shell disease was seen in the small shrimp *Crangon septemspinosa*, an important estuarine and coastal food chain organism common on the east coast of North America, where it is a part of the diet of bluefish, weakfish, flounders, sea bass, and other economic species. Examination of samples of *Crangon* from the New York Bight disclosed high prevalences (up to 15%) of eroded appendages and blackened erosions of the exoskeleton. The disease condition was only rarely observed at less degraded collecting sites (e.g., Beaufort, North Carolina, and Woods Hole, Massachusetts). Histological examination of diseased specimens produced findings similar to those with crabs and lobsters. All layers of the exoskeleton were eroded; affected portions were brittle and easily fragmented; cracking and pitting of calcified layers occurred; and underlying tissues were often necrotic. Laboratory experiments using seawater from the highly polluted inner New York Bight resulted in the appearance of the disease in 50% of individuals. Shell erosion was progressive; crippled individuals were cannibalized; and eroded segments of appendages did not regenerate after molting. No signs of disease developed in control animals held in artificial seawater.

During the same period, a German study of the effects of industrial wastes on the brown shrimp, *Crangon crangon,* disclosed a high prevalence of so-called black spot disease, with signs similar to those seen in *C. septemspinosa* from the New York Bight.[6] Juvenile and adult shrimp from the polluted Fohr River estuary in northwestern Germany had black areas of erosion on the carapace and the appendages, with necrosis of the underlying tissues, and, frequently, missing terminal segments of appendages. The disease condition varied seasonally in prevalence, with a peak of 8.9% in the summer. Lesions persisted and worsened after molting.

A recent study of shell disease and the related black gill disease in lobsters from Massachusetts waters showed similar trends for both diseases, with the highest prevalences in samples from the most polluted sites — particularly Boston Harbor and Buzzards Bay.[7] Mortalities were not observed, but population impacts were considered likely.

Shell disease and black gill disease were also examined recently in rock crabs, *Cancer irroratus,* from the Atlantic coast of North America.[8] High prevalences were associated with sewage and/or dredge spoil dumping areas, where environmental degradation was worst. Prevalences of black gill disease were as high as 30% in some samples from the New York Bight apex. The etiology of the gill blackening condition is clearly complex, but it involves the presence of black silt between gill lamellae as well as the presence of fouling organisms. Foci of black discoloration were found to be areas in which several adjacent gill lamellae were dead and necrotic, with accompanying blackening and necrosis of the gill tissue.

A detailed review published in 1989 of shell disease in crustaceans of commercial importance[9] resulted in the following general conclusions about relationships with pollution:

1. Shell disease may occur with higher prevalence and greater severity in polluted areas than in those not degraded by man's activities. The balance between metabolic processes associated with new shell formation, and infection by microbes capable of utilizing chitin, may be disturbed by environmental changes affecting normal shell formation or favoring the growth of chitin-utilizing microbes. Such disturbances may be consequences of pollution.
2. Evidence exists for an association of shell disease with habitat degradation. Prevalences have been found to be high in crustaceans from polluted sites; prevalences show trends similar to those of the black gill syndrome, which also has a statistical association with extent of pollution. Experimental exposures of crustaceans to contaminated sediments, heavy metals, biocides, petroleum, and petroleum derivatives can result in the appearance of the black gill syndrome, often accompanied by shell disease.

The physiology of crustaceans, especially that related to hormonal control of molting, has been elucidated by many studies during the past half-century. The role of pollutants in altering metabolic pathways involved is of course an area of concern. As an example, abnormal production of the steroid-molting hormones may inhibit shell synthesis, whereas hormonal insufficiency may delay or prevent molting, thus affecting growth and survival. These metabolic anomalies may enhance the effects of shell disease, especially when accompanied by diminution of other cellular and humoral mechanisms of internal defense. The significance of an enzyme (prophenoloxidase-activating) system in crustacean responses to infection has been pointed out in a series of recent papers,[10] and the functioning of this system in the presence of pollutants can be important to host survival.

The appearance of shell erosion may therefore be the consequence of a disturbed balance between processes of chitin maintenance and repair and the activities of chitinoclastic microorganisms — this disturbance created by either natural or man-made environmental changes. Critical to an understanding of the relationship are environmental, genetic, and immunological factors that may either promote repair or, conversely, enhance exoskeletal degradation.

Activated Latent Infections

Recently published accounts of two viral diseases of marine invertebrates indicate that latent infections may be provoked into patency by environmental stress. One, a *Baculovirus* infection of pink shrimp, *Penaeus duorarum,* was first recognized in stressed laboratory populations.[11] The other, a herpes-like viral infection of oysters, was discovered in a population held in a heated power plant effluent in Maine.[12]

The relationship of a shrimp viral disease and chronic exposure to pollutant chemicals has been explored at the Gulf Breeze (FL) Environmental Research Laboratory of the U.S. Environmental Protection Agency. In this research, a viral disease of Gulf of Mexico pink shrimp caused by *Baculovirus penaei* reached patent levels and caused mortalities of 50 to 80% in shrimp exposed to the PCB Aroclor 1254 and the organochlorine insecticide Mirex. Other experiments in which the shrimp were crowded, but not exposed to chemicals, resulted in similar enhancement of viral infections, indicating that environmental stress may be an important determinant of patent infections and death. The viral infections have been found subsequently in brown and white shrimp of the Gulf of Mexico as well.

An association of high environmental temperatures with high disease prevalence (or disease enhancement) in molluscan shellfish sampled from thermal effluents has been made recently. A lethal herpes-type viral disease was reported in oysters, *Crassostrea virginica,* held in heated discharge water from an electric generating station in Maine. The disease, which apparently existed at a low level in oysters growing at normal, low environmental temperatures (12 to 18°C summer temperatures), proliferated in oysters maintained at the elevated temperatures (28 to 30°C), and produced mortalities in those stressed populations. Intranuclear inclusion bodies containing viral particles characterized advanced infections. Mortalities of oysters held at higher temperatures were correlated with a greater prevalence of the viral inclusions. The elevated temperatures were considered by the investigators to favor spread of the infection or to activate latent infections, or both.

Reduced Disease Resistance

Reduction of shellfish disease resistance caused by contaminant exposure has been suggested in the previous sections on crustacean shell disease and shrimp viral disease. Additional experimental evidence exists as well. One study exposed hard clams, *Mercenaria mercenaria,* to phenol, and found damage to gill and digestive tract epithelia — tissues that are considered to be important components of internal defense systems.[13] The investigators suggested, but did not demonstrate, that phenol-treated clams might be more susceptible to microbial infections than normal ones. A mechanism for increased susceptibility of clams to bacterial infection was proposed in another study[14] based on chronic (18-week) exposure to benzo[*a*]pyrene (BaP), pentachlorophenol, and hexachlorobenzene. Most of the exposed clams showed significantly impaired ability to clear injected bacteria, indicating that resistance to bacterial invasion was decreased by exposure to pollutant chemicals. The deficiency in bacterial clearance was directly proportional to tissue levels of the chemicals used (some clams with high tissue burdens were unable to clear bacteria). Bacterial clearance may result from the activity of cellular and humoral factors, and a number of these have been demonstrated in hard clams.[15] Phagocytic activity was reduced after exposure to pollutants, as was the presence of bacterial agglutinins and lysins.

Pollutants may affect the cellular defenses of molluscs in a positive way as well.[16] Phagocytic activity of oyster blood cells (hemocytes) has been reported to be enhanced by exposure to heavy metals. In studies of mussels exposed to gradually increasing levels of copper ions, a dramatic increase in circulating granulocytes was seen, as was an increase in hemolymph lytic enzymes, total protein, and proteolytic activity. Binding and sequestering of excess copper was a later phenomenon. In other studies with invertebrates, environmental stress increased blood glucose levels in lobsters, *Homarus americanus,* and crayfish, *Cambarus clarkii.*

Generally, though, there is little information available about the effects of toxic chemicals on disease resistance in molluscs and crustaceans. Reaching out to other phyla, exposure of a polychaete, *Glycera dibranchiata,* to pentachlorophenol and hexachlorobenzene induced release of a bactericidal protein, but decreased the effectiveness of phagocytes.[17] The investigators suggested that the release of the bactericidal protein from coelomocytes was similar to macrophage activation in vertebrates and that the recognition and binding mechanisms of phagocytes were the components affected by exposure to pollutants (possibly leading to the decreased ability to recognize and destroy pathogens). Humoral resistance factors, known to exist in the worm, were unaffected by the pollutant dosages used.

Some experimental evidence exists to suggest a causal relationship between specific pollutant chemicals and fungal parasitization of shellfish. In one study,[18] oysters exposed to pesticides (e.g., DDT, toxaphene, and parathion) became infected with a mycelial fungus (not further identified) that caused lysis of the mantle, gut, gonads, gills, visceral ganglion, and kidney tubules. None of the control oysters became infected, indicating a role for the pesticides in altering the host–parasite relationship of the oysters and the fungus.

Results of other investigations also suggest that chronic pollution pressure may place a stress on the animal that could be manifested in increased parasitism.[19] Mussels, *Mytilus edulis*, from polluted areas of the northeast coast of the United States were more extensively and intensively parasitized (three to five species of unidentified parasites) than were those from other areas. The investigators pointed out that the increased parasitism occurred in waters near centers of human population density and industrial activity. In a related study of the effects of chronic exposure to various crude oils on pond-held oysters, a similar increase in the prevalence and intensity of parasites was found — with the conclusion that resistance to infection can be reduced in polluted environments, increasing susceptibility to some parasites.

An important defensive response of crustaceans to pollutants has been labeled "the black gill syndrome" (Figure 22), which seems to be a significant blood cell (hemocyte)-mediated defense response of Crustacea to chemical or physical trauma or to pathogen invasion. It results in the sequestering and subsequent removal during molting of toxic substances or microbial invaders. It also can result in truncated gill filaments, a reduced surface area, and consequent reduction in gas and ion exchanges.

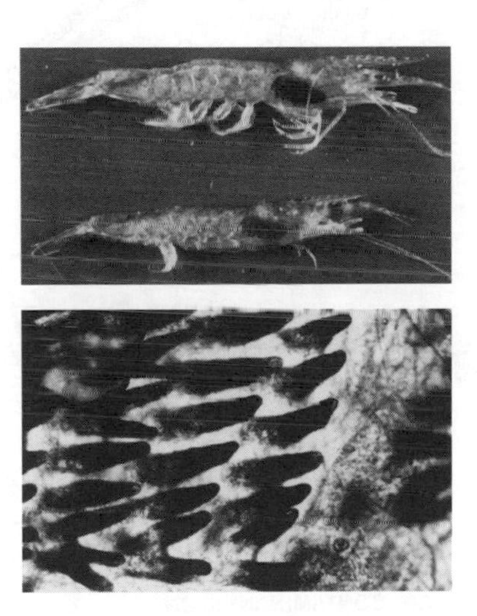

FIGURE 22. Black gills in juvenile shrimp. Shown are the gross appearance *(top)* and a wet mount of a gill process *(bottom)*. (Photographs courtesy of Dr. D.V. Lightner.)

The functional aspect of the syndrome is the formation of apical cellular plugs consisting of hemocytes formed in response to gill trauma or systemic microbial infection. Phagocytized material is aggregated in gill apices and eliminated during molting. Gill blackening has been reported as a consequence of protozoan and fungal infection and cadmium exposure.[20] In one study, crustacean defensive responses to toxicant exposure were found to consist of encapsulation by circulating hemocytes of degenerated tissues resulting from the action of toxicants, followed by their concentration in apices of gill lamellae.[21] These responses, or their extrapolations, may be important to the survival of crustaceans in degraded habitats and may be altered by chemicals or other stressors.

POLLUTION AND NONINFECTIOUS DISEASES

The range of noninfectious diseases of shellfish — those abnormalities that cannot be specifically related to activities of a particular pathogen — is broad, encompassing numerous biochemical/physiological defects, including those that may result in structural changes in tissues of the animals. Toxic pollutants may of course be initiators of some of the observed physiological or structural abnormalities, usually by interfering with normal cell metabolic pathways or by inducing genetic changes in eggs or larvae (Figure 23).

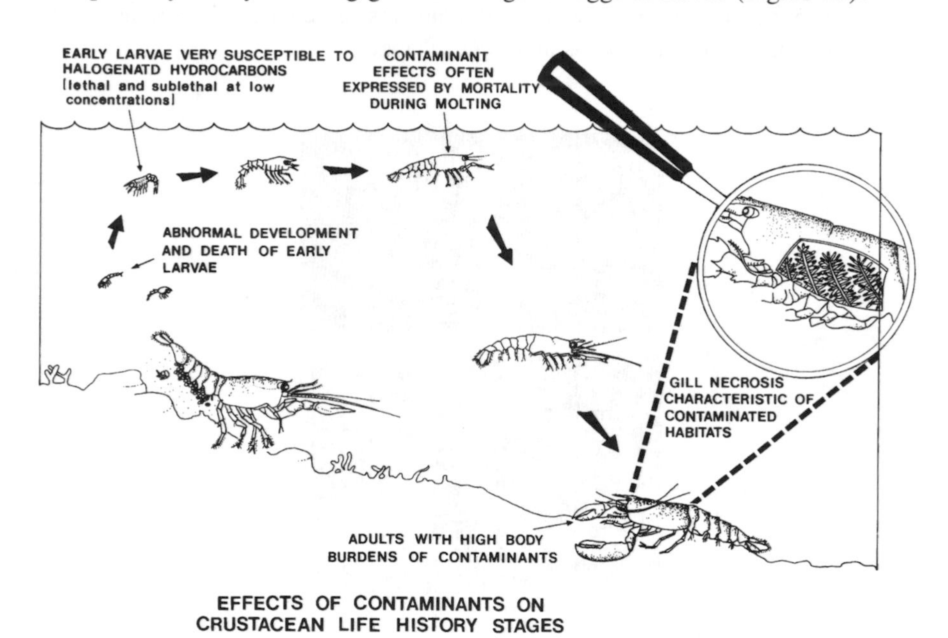

FIGURE 23. Contaminant effects on crustacean life cycle stages.

Pollution-associated noninfectious diseases of shellfish to be included here can be categorized generally as cellular damage and mutations caused by pollutants, shell abnormalities in oysters, and neoplasms (cancers) in molluscs.

Cellular Damage and Mutations Caused by Pollutants

The mutation-inducing (mutagenic) properties of a number of chemical contaminants, including heavy metals, pesticides, and petroleum-derived polycyclic hydrocarbons, have been demonstrated repeatedly in experimental studies with terrestrial animals, and it is likely that at least some marine organisms will respond to chemical mutagens in species-specific ways, but probably with differing sensitivities.

Several assay systems exist for detecting the presence of mutagens in the marine environment. Using one such system, mussels were sampled from polluted and unpolluted waters of the United Kingdom, and extracts of their tissues were tested for the ability to induce genetic changes in bacterial and yeast cultures.[22] Significant increases in mutation rates for specific gene loci were seen in cultures exposed to extracts of mussels from polluted waters, but not in those from clean waters, providing evidence for the presence of mutagens in the tissues of mussels from the contaminated areas. The chemical nature of the mutagens was not identified in the study, except that the mussels came from areas with heavy industrial pollution.

Other microbial assay systems for the detection of environmental mutagens and carcinogens have been developed.[23] Many are based on a bacterial test for mutagens described two decades ago; several could have relevance to the assessment of pollutants in marine animals. One of the most promising assay approaches is the "sister chromatid exchange" technique, which has been used to measure mutagenicity in mussels exposed to environmental contaminants. The degree of chromosomal damage is measured by differential staining properties and is an indicator of mutagen concentration.

Looking beyond the process to the results of exposing developing embryos to pollutants, several studies have made use of commercial shellfish species, including oysters, *Crassostrea virginica,* and periwinkles, *Littorina saxatilis*.[24] Heavy metal toxicity to oyster embryos has been described, and a study of developmental abnormalities in the embryos of periwinkles disclosed a direct relationship with the degree of pollution stress to which the adult population had been exposed. Samples from polluted sites produced much higher prevalences of deformed embryos than did those from a clean site.

Exposure of oysters, *C. virginica,* to the carcinogens benzo[a]pyrene (BaP) and 3-methylcholanthrene resulted in inflammation of the connective tissue surrounding blood vessels and proliferation of a cell type closely resembling those found in neoplasms of natural populations of oysters and other bivalves (to be described later in this chapter).[25] The abnormal cells exhibited mitotic activity and were larger than normal oyster hemocytes; these are predominant characteristics of bivalve neoplastic cells. The investigators did not, however, consider the lesions to be unequivocal neoplasms, and the experimentally induced lesions were not as extensive as those seen in wild populations of bivalves.

Shell Abnormalities in Oysters

Pacific oysters, *Crassostrea gigas,* were introduced to the coast of France in the late 1960s and to the east coast of Britain in the early 1970s. By the mid-1970s, a serious problem with reduced growth and malformed shells was apparent, and the cause was found to be organotin compounds used on boats in antifouling paints. Abnormally thickened shells in exposed populations of oysters had an open laminar structure, a formation described as "chambering," sometimes with a proteinaceous gel secreted between the layers. In one French study, such abnormalities in shell calcification reached 90% during 1980 to 1982 in the Bay of Arcachon, but fell to 40% in 1983 to 1985, after a ban on the use of tributyltin in antifouling paint was imposed.[26] The severity of the deformations and the tissue levels of tin also decreased during the same period, confirming the chemical etiology of the condition. The extreme sensitivity of oysters to tributyltin is apparent from findings that concentrations of only 50 ppb could induce shell malformations. In a British study,[27] *C. gigas* spat cultured in the presence of low environmentally feasible levels of the contaminant grew less well than controls and developed pronounced thickening and chambering of the upper shell valve. Higher levels of exposure resulted in increasing percentages of mortalities. It seems relevant to note here that in other studies,[28] experimental exposure of crabs, *Uca pugilator,* to tributyltin retarded limb regeneration, delayed ecdysis, and produced deformities in regenerated appendages. Decreased survival and reduced growth of lobster and crab larvae were also observed following exposure to tributyltin.

Neoplasms (Cancers) in Molluscs

Although cancer has been studied most extensively in humans and in laboratory mammals, the existence of tumors in shellfish has been recognized for more than a century (the first oyster tumor, for example, was reported in 1887).[29] The possible role of environmental chemical factors in inducing neoplasms in shellfish is uncertain, but a growing body of information is becoming available.

Neoplastic growths in the softshell clam, *Mya arenaria,* were examined in the early 1970s in relation to petroleum contamination.[30] Gonadal and hematopoietic (blood-forming tissue) neoplasms were observed in animals collected from two chronically contaminated sites on the Maine coast, with the finding of prevalences up to 29% in some samples. The investigators (M. M. Barry and P. P. Yevich) stated that "no tumors similar to those described [from the petroleum contaminated area] have been encountered in animals collected from any other area." They described the scope of their study as "several thousand animals from all coastal areas of the United States."

Then, beginning in 1976 a greatly expanded study of the prevalences of neoplasms in softshell clams was instituted to determine whether the abnormalities were associated with varying types and degrees of environmental hydrocarbon pollution.[31] Ten coastal sites from Maine to Rhode Island were selected, based on the existence of various kinds and intensities of pollution. Examination of 1,829 clams disclosed neoplasia in 159 (8%), but at only 5 of the 10 sites. Prevalences varied drastically from site to site — from 0 to 64%. Gonadal and hematopoietic neoplasias were found, and those of presumed hematopoietic origin were observed to cause poor condition and mortality. An interesting geographic variation was seen, with only hematopoietic neoplasms found in southern New England and predominantly gonadal neoplasms found in northern New England. Unfortunately, the relationship of clam neoplasms to oil pollution was not clarified by the study. In the words of the investigators (R. S. Brown and colleagues), "There is a surprising dichotomy of results from petroleum-derived hydrocarbon (PDH)-polluted sites. Clams from some polluted sites had no neoplasms, whereas other PDH-polluted sites had a high prevalence of neoplasms. These results suggested that the type and degree of hydrocarbon pollution are possibly related to the frequency of neoplasms and other lesions in *Mya,* but they are by no means the only causative factors." The investigators went on to state that "the possibility that neoplastic disease in *Mya* has a genetic base or *is a result of an infectious agent* cannot be excluded by our results" (emphasis added).

Then, in 1979 and 1980, experimental induction of the hematopoietic neoplasms in clams was reported.[32] Exposure of normal clams to water that had passed over populations of neoplastic clams, with or without the presence of petroleum hydrocarbons, resulted in significant levels of neoplasia in experimental animals. Prevalence in one experimental exposure exceeded 70%. This was the first reported transmission of neoplasia in marine invertebrates, and it reinforced the likelihood of an infectious process.

Another chapter in the softshell clam neoplasia story was published in 1981. A viral etiological agent was described and was reported to be similar to β-type retroviruses.[33] Purified virus, when injected into normal clams, was reported to cause tumors within two months. No direct correlation between disease prevalence and the type or extent of environmental pollution was demonstrated in any of the experiments, but the role of petroleum hydrocarbons as stressors was not discounted.

There are many other reports of neoplasms in molluscs.[34] Hematopoietic neoplasms have been observed in a number of species in addition to *Mya arenaria*. Gonadal neoplasms have been reported in quahogs, *Mercenaria mercenaria,* from Narragansett Bay, Rhode Island. Samples collected in 1968, 1969, and 1970 had tumor frequencies of 0.2, 2.3, and 2.7%, respectively. In another study, epizootic neoplasms (up to 10% prevalence) were found in a localized population of the clam *Macoma balthica* from a tributary of Chesapeake Bay. The neoplasms were invasive and systemic, with initial foci in the gill epithelia. Holding experiments indicated that the disease was usually fatal. No association with environmental factors was made.

Epizootic levels of neoplasms with a possible environmental etiology were reported in several molluscan species from Yaquina Bay, Oregon.[35] Blue mussels, *Mytilus edulis,* native oysters, *Ostrea lurida,* and two species of *Macoma* were affected, and winter mortalities were associated with the disease. In a related chemical study, mussels from polluted Oregon estuaries were found to have significant body benzo[a]pyrene (BaP) burdens and neoplasms, while those with low or undetectable levels of BaP had no neoplasms. However, there have been other reports from Oregon that failed to disclose any association of molluscan neoplasia and pollution.

Mussels, *Mytilus edulis,* from British waters were also found to have a "proliferative atypical hemocytic condition".[36] Described as "infiltration and replacement of connective tissue by enlarged, atypical, mitotically active, basophilic, hemocyte-like cells," the disorder was seen in varying degrees of severity. Advanced cases were characterized by extensive replacement of connective tissue and degeneration of digestive gland cells. The condition was remarkably similar to that observed in mussels from Oregon, and samples from a hydrocarbon-polluted site in southern England (Plymouth) disclosed a prevalence of 1.6%. Sediments from the collection site contained several aromatic hydrocarbons, but the investigators stated that "any speculation about the etiology of the [proliferative] condition described must remain inconclusive."

CONCLUSIONS

Based on these few examples of pollution-associated diseases of shellfish, a number of tentative conclusions seem warranted:

1. Environmental stress from pollutants seems to be an important determining factor in several shellfish diseases. Effects include direct chemical damage to membranes or tissues, modification of physiological and biochemical pathways, increased infection pressure from facultative microbial pathogens, and reduced resistance to infection.
2. The presence of marginal or degraded coastal/estuarine environments may be signaled by the appearance, or the increase in prevalence, of several diseases, including shell disease in crustaceans and certain neoplasms in bivalve molluscs, but a clear cause-and-effect relationship has not yet been demonstrated for most of these diseases.
3. The multifactorial genesis of disease in marine species is becoming apparent, involving environmental stress, facultative pathogens, resistance of hosts, and latent infections.
4. A number of viruses have been found in crustaceans and molluscs in recent years, and the pathogenic role of two of them (shrimp *Baculovirus* and oyster herpes-like virus has been demonstrated by exposures to increasing environmental stress. Other latent viral infections of invertebrates may be identified by similar experimental methods.
5. Among the most severe and persistent problems in establishing pollutant–disease relationships are (a) the absence of baseline information about the organisms and their habitats prior to pollution, (b) the existence of multiple pollutants in most heavily degraded habitats, and (c) the circumstantial (and still controversial) nature of much of the evidence linking pollution and disease.

6. Experimental exposure of crustaceans to toxic chemicals has provided data useful to understanding defense mechanisms, especially the role of the black gill syndrome in sequestering and then removing toxic substances and pathogens.

In concluding this chapter, it might be relevant — just for balance — to summarize opinions on the subject of pollution-associated diseases published in 1985 by the International Council for the Exploration of the Sea's Working Group on Pathology and Diseases of Marine Organisms. Principal conclusions (with which I do not necessarily agree) were:

1. An established link between diseases and pollution does not exist.
2. While environmental factors have a role in the etiology of at least some diseases, most diseases have a multifactorial etiology.
3. Environmental contaminants do not always have the same effect on different species; even when chemicals cause anomalies in one marine population, anomalies may not be observed in another population exposed to the same chemicals.
4. It may be possible, however, to use certain pathological signs as indicators of general water quality deterioration.
5. The most likely general effect of pollution is a nonspecific lowering of resistance, resulting in the appearance of variable disease signs rather than the specific induction of an identifiable syndrome.

These opinions of the ICES Working Group seem unnecessarily cautious and negative, however, in view of the information summarized in this chapter and the preceding one on fish diseases. Whereas it is certainly true that some of the evidence is circumstantial and that more data are needed, it seems nonetheless that a considerable measure of association between certain fish and shellfish diseases and pollution has been demonstrated. Studies using a combination of field observations and experimental exposures to contaminants have done much and could do much more to reduce uncertainties still further.

Despite the *relative* lack of robustness of the data base, it is becoming apparent that many of the disorders of molluscs and crustaceans in degraded habitats are stress related and are often expressions of a stress syndrome, quite distinct from that of fish, yet similar in that heterostasis — a heightened, energy-demanding physiological response to continuing stress — is achieved.

REFERENCES

1. **Jeffries, H. P.** 1972. A stress syndrome in the hard clam, *Mercenaria mercenaria*. *J. Invertebr. Pathol*. 20: 242–251.
2. **Bayne, B. L.** 1975. Aspects of physiological condition in *Mytilus edulis* L., with special reference to the effects of oxygen tension and salinity. *Proc. 9th Eur. Mar. Biol. Symp*., pp. 213–238; **Bayne, B. L., M. N. Moore, J. Widdows, D. R. Livingstone, and P. Salkeld.** 1979. Measurement of the response of individuals to environmental stress and pollution: studies with bivalve molluscs. *Philos. Trans. R. Soc. Lond*., Ser. B 286: 563–581.

3. **Sindermann, C. J.** 1989. The shell disease syndrome in marine crustaceans. U.S. Department of Commerce, National Marine and Fishery Service, NOAA Tech. Memo. NMFS-F/NEC-64, 43 pp.
4. **Pearce, J. B. (Ed.).** 1971. The effects of waste disposal in the New York Bight — final report (draft) for 9 July 1971 to Coastal Engineering Research Center, U.S. Corps of Engineers, Washington, D.C. Prepared by Sandy Hook Sport Fish. Mar. Lab., Highlands, NJ; **Pearce, J. B.** 1972. The effects of solid waste disposal on benthic communities in the New York Bight, pp. 404–411. in Ruivo, M. (Ed.), *Marine Pollution and Sea Life*. Fish News Ltd., London; **Young, J. S. and J. B. Pearce.** 1975. Shell disease in crabs and lobsters from New York Bight. *Mar. Pollut. Bull.* 6: 101–105.
5. **Gopalan, U. K. and J. S. Young.** 1975. Incidence of shell disease in shrimps in the New York Bight. *Mar. Pollut. Bull.* 6: 149–153.
6. **Schlotfeldt, H. J.** 1972. Jahreszeitliche Abhängigkeit der "Schwarzflecken-krankheit" bei der Garnele, *Crangon crangon* L. *Ber. Dtsch. Wiss. Komm. Meerseforsch.* 22: 397–399.
7. **Estrella, B. T.** 1984. Black gill and shell disease in American lobster *(Homarus americanus)* as indicators of pollution in Massachusetts Bay and Buzzards Bay, Massachusetts. Mass. Dept. Fish. Wildl. Rec. Veh., Div. Mar. Fish. Publ. 14049–19–125–5-85-C.R., 17 pp.
8. **Sawyer, T. K.** 1982. Distribution and seasonal incidence of "black gill" in the rock crab, *Cancer irroratus,* pp. 199–211. in Mayer, G. F. (Ed.), *Ecological Stress and the New York Bight: Science and Management*. Estuar. Res. Fed., Columbia, SC; **Sawyer, T. K., E. J. Lewis, M. Galasso, S. Bodammer, J. Ziskowski, D. Lear, M. O'Malley, and S. Smith.** 1983. Black gill conditions in the rock crab *Cancer irroratus,* as indicators of ocean dumping in Atlantic coastal waters of the United States. *Rapp. P.-V. Reun., Cons. Int. Explor. Mer* 182: 91–95; **Sawyer, T. K., E. J. Lewis, M. E. Galasso, and J. J. Ziskowski.** 1984. Gill fouling and parasitism in the rock crab, *Cancer irroratus* Say. *Mar. Environ. Res.* 14: 355–369.
9. **Sindermann, C. J., F. Csulak, T. K. Sawyer, R. A. Bullis, D. W. Engel, B. T. Estrella, E. J. Noga, J. B. Pearce, J. C. Rugg, R. Runyon, J. A. Tiedemann, and R. R. Young.** 1989. Shell disease of crustaceans in the New York Bight. U.S. Department of Commerce, NOAA Tech. Memo. NMFS-F/NEC-74, 47 pp.
10. **Söderhäll, K.** 1982. Prophenoloxidase activating system and melanization — a recognition mechanism of arthropods? A review. *Dev. Comp. Immunol.* 6: 601–611; **Söderhäll, K.** 1983. 1,3-glucan enhancement of protease activity in crayfish haemocyte lysate. *Comp. Biochem. Physiol.* B 74: 221–224; **Söderhäll, K.** 1986. The cellular immune system in crustaceans, pp. 417–420. in Samson, R. A., J. M. Vlak, and D. Peters (Eds.), *Fundamental and Applied Aspects of Invertebrate Pathology*. Found. Fourth Int. Colloq. Invertebr. Pathol., Wageningen, The Netherlands; **Söderhäll, K. and R. Ajaxon.** 1982. Effect of quinones and melanin on mycelial growth of *Aphanomyces* spp. and extracellular protease of *Aphanomyces astaci* parasite on crayfish. *J. Invertebr. Pathol.* 39: 105–109; **Söderhäll, K. and V. J. Smith.** 1983. Separation of the haemocyte populations of *Carcinus maenas* and other marine decapods and prophenoloxidase distribution. *Dev. Comp. Immunol.* 7: 229–239; **Söderhäll,**

K. and L. Häll. 1984. Lipopolysaccharide induced activation of prophenoloxidase activating system in crayfish haemocyte lysate. *Biochim. Biophys. Acta* 797: 99–104; **Söderhäll, K. and V. J. Smith.** 1986. The prophenoloxidase activating system: the biochemistry of its activation and role in arthropod cellular immunity with special reference to crustaceans, pp. 208–223. in Brehelin, M. (Ed.), *Immunity in Invertebrates*. Springer-Verlag, Berlin; **Söderhäll, K., A. Wingren, M. W. Johansson, and K. Bertheussen.** 1985. The cytotoxic reaction of hemocytes from the freshwater crayfish, *Astacus astacus*. *Cell. Immunol.* 94: 326–330.

11. **Couch, J. A.** 1974a. Free and occluded virus, similar to *Baculovirus*, in hepatopancreas of pink shrimp. *Nature* (Lond.) 247: 229–231; **Couch, J. A.** 1974b. An enzootic nuclear polyhedrosis virus of pink shrimp: ultrastructure, prevalence, and enhancement. *J. Invertebr. Pathol.* 24: 311–331; **Couch, J. A. and D. R. Nimmo.** 1974a. Ultrastructural studies of shrimp exposed to the pollutant chemical, polychlorinated biphenyl (Aroclor 1254). *Bull. Pharmacol. Environ. Pathol.* 11: 17–20; **Couch, J. A. and D. R. Nimmo.** 1974b. Detection of interactions between natural pathogens and pollutants in aquatic animals. Ctr. Wetl. Resour. Publ. LSU-SG-74-05, Louisiana State University, Baton Rouge, LA, pp. 261–268.
12. **Farley, C. A., W. G. Banfield, G. Kasnick, Jr., and W. S. Foster.** 1972. Oyster herpes-type virus. *Science* 178: 759–760.
13. **Fries, C. and M. R. Tripp.** 1976. Effects of phenol on clams. *Mar. Fish. Rev.* 38(10): 10–11.
14. **Anderson, R. S., C. S. Giam, L. E. Ray, and M. R. Tripp.** 1981. Effects of environmental pollutants on immunological competency of the clam *Mercenaria mercenaria:* impaired bacterial clearance. *Aquat. Toxicol.* 1: 187–195.
15. **Arimoto, R. and M. R. Tripp.** 1977. Characterization of a bacterial agglutinin in the hemolymph of the hard clam, *Mercenaria mercenaria*. *J. Invertebr. Pathol.* 30: 406–413; **Fries, C. R. and M. R. Tripp.** 1980. Depression of phagocytosis in *Mercenaria* following chemical stress. *Dev. Comp. Immunol.* 4: 233–244.
16. **Cheng, T. C. and J. T. Sullivan.** 1984. Effects of heavy metals on phagocytosis by molluscan hemocytes. *Mar. Environ. Res.* 14: 305–315; **Pickwell, G. V. and S. A. Steinert.** 1984. Serum biochemical and cellular responses to experimental cupric ion challenge in mussels. *Mar. Environ. Res.* 14: 245–265.
17. **Anderson, R. S., C. S. Giam, and L. E. Ray.** 1984. Effects of hexachlorobenzene and pentachlorophenol on cellular and humoral immune parameters in *Glycera dibranchiata*. *Mar. Environ. Res.* 14: 317–326.
18. **Lowe, J. I., P. D. Wilson, A. J. Rick, and A. J. Wilson, Jr.** 1971. Chronic exposure of oysters to DDT, toxaphene and parathion. *Proc. Natl. Shellfish Assoc.* 61: 71–79.
19. **Yevich, P. P. and C. A. Barszcz.** 1983. Histopathology as a monitor for marine pollution. Results of histopathological examinations of the animals collected for the US 1976 Mussel Watch Program. *Rapp. P.-V. Reun., Cons. Int. Explor. Mer* 182: 96–102; **Barszcz, C. A., P. P. Yevich, L. R. Brown, J. D. Yarbrough, and C. D. Minchew.** 1978. Chronic effects of three crude oils on oysters suspended in estuarine ponds. *J. Environ. Pathol. Toxicol.* 1: 879–895.

20. **Couch, J. A.** 1978. Diseases, parasites, and toxic responses of commercial penaeid shrimps of the Gulf of Mexico and South Atlantic coasts of North America. *Fish. Bull.* 76: 1–44; **Lightner, D. V.** 1975. Some potentially serious disease problems in the culture of penaeid shrimp in North America. *Proc. U.S.–Japan Nat. Resour. Prog.: Symp. Aquacult. Dis.* 1974, pp. 75–97.
21. **Doughtie, D. G. and K. Rao.** 1983. Ultrastructural and histological study of degenerative changes leading to black gills in grass shrimp exposed to a dithiocarbamate biocide. *J. Invertebr. Pathol.* 41: 33–50.
22. **Parry, J. M., D. J. Tweats, and M. A. Al-Mossawi.** 1976. Monitoring the marine environment for mutagens. *Nature* (London) 264: 538–540.
23. **Ames, B. N.** 1972. A bacterial system for detecting mutagens and carcinogens, pp. 57–66. in Sutton, H.E. and M.I. Harris (Eds.), *Mutagenic Effects of Environmental Contaminants*. Academic Press, New York; **Dixon, D. R.** 1983. Sister chromatid exchange and mutagens in the aquatic environment. *Mar. Pollut. Bull.* 14: 282–284; **Dixon, D. R. and K. R. Clarke.** 1982. Sister chromatid exchange, a sensitive method for detecting damage caused by exposure to environmental mutagens in the chromosomes of adult *Mytilus edulis*. *Mar. Biol. Lett.* 3: 163–172.
24. **Calabrese, A., R. S. Collier, D. A. Nelson, and J. R. MacInnes.** 1973. The toxicity of heavy metals to embryos of the American oyster *Crassostrea virginica*. *Mar. Biol.* (Berl.) 18: 162–166; **Dixon, D. R. and D. Pollard.** 1985. Embryo abnormalities in the periwinkle, *Littorina "saxitilis,"* as indicators of stress in polluted marine environments. *Mar. Pollut. Bull.* 16: 29–33.
25. **Couch, J. A., L. A. Courtney, J. T. Winstead, and S. S. Foss.** 1979. The American oyster *(Crassostrea virginica)* as an indicator of carcinogens in the aquatic environment, pp. 65–84. in *Animals as Monitors of Environmental Pollutants*. National Academy of Science, Washington, D.C.
26. **Alzieu, C., M. Héral, Y. Thibaud, M. J. Dardignac, and M. Feuillet.** 1982. Influence des peintures antisalissures à base d'organostanniques sur la calcification de la coquille de l'huître *Crassostrea gigas*. *Rev. Trav. Inst. Peches Marit.* 44: 301–349; **Alzieu, C., J. Sairjuan, J.-P. Deltreil, and M. Borel.** 1986. Tin contamination in Arcachon Bay: effects on oyster shell anomalies. *Mar. Pollut. Bull.* 17: 494–498.
27. **Waldock, M. J. and J. E. Thain.** 1983. Shell thickening in *Crassostrea gigas:* organotin antifouling or sediment induced? *Mar. Pollut. Bull.* 14: 411–415.
28. **Weis, J. S., J. Gottlieb, and J. Kwiatkowski.** 1987. Tributyltin retards regeneration and produces deformities of limbs in the fiddler crab, *Uca pugilator*. *Arch. Environ. Contam. Toxicol.* 16: 321–326; **Laughlin, R. B. and W. J. French.** 1980. Comparative study of the acute toxicity of a homologous series of trialkyltins to larval shore crabs, *Hemigrapsus nudus,* and lobster, *Homarus americanus*. *Bull. Environ. Contam. Toxicol.* 25: 802–809.
29. **Ryder, J. A.** 1887. On a tumor in oyster. *Proc. Natl. Acad. Sci. U.S.A.* 44: 25–27.
30. **Barry, M. M. and P. P. Yevich.** 1972. Incidence of gonadal cancer in the quahog *(Mercenaria mercenaria)*. *Oncology* 26: 87–96; **Barry, M. M. and P. P. Yevich.** 1975. The ecological, chemical and histopathological evaluation of an oil spill site. III. Histopathological studies. *Mar. Pollut. Bull.* 6: 171–173; **Yevich, P. P. and C. A. Barszcz.** 1976. Gonadal and hematopoietic neoplasms

in *Mya arenaria*. *Mar. Fish. Rev.* 39(10): 42–43; **Yevich, P. P. and C. A. Barszcz.** 1977. Neoplasia in soft-shell clams *(Mya arenaria)* collected from oil-impacted sites. *Ann. N.Y. Acad. Sci.* 298: 409–426.

31. **Brown, R. S., R. E. Wolke, and S. B. Saila.** 1976. A preliminary report on neoplasia in feral populations of the soft-shell clam, *Mya arenaria:* prevalence, histopathology and diagnosis. *Proc. Int. Colloq. Invertebr. Pathol.,* pp. 151–159; **Brown, R. S., R. E. Wolke, S. B. Saila, and C. W. Brown.** 1977. Prevalence of neoplasia in 10 New England populations of the soft-shell clam *(Mya arenaria). Ann. N.Y. Acad. Sci.* 298: 522–534; **Walker, H. A., E. Lorda, and S. B. Saila.** 1981. A comparison of incidence of five pathological conditions in soft shell clams *Mya arenaria* from environments with various pollution histories. *Mar. Environ. Res.* 5: 109–123.
32. **Brown, R. S., R. E. Wolke, C. W. Brown, and S. B. Saila.** 1979. Hydrocarbon pollution and the prevalence of neoplasia in New England soft-shell clams *(Mya arenaria),* pp. 41–51. in *Animals as Monitors of Environmental Pollutants*. National Academy of Science, Washington, D.C; **Brown, R. S.** 1980. The value of the multidisciplinary approach to research on marine pollution effects as evidenced in a three-year study to determine the etiology and pathogenesis of neoplasia in the soft-shell clam, *Mya arenaria. Rapp. P.-V. Reun., Cons. Int. Explor. Mer* 179: 125–128; **Appeldoorn, R. S. and J. J. Oprandy.** 1980. Tumors in soft-shell clams and the role played by a virus. *Maritimes* 24(1): 4–6.
33. **Oprandy, J. J., P. W. Chang, A. D. Pronovost, K. R. Cooper, R. S. Brown, and V. J. Yates.** 1981. Isolation of a viral agent causing hematopoietic neoplasia in the soft-shell clam, *Mya arenaria. J. Invertebr. Pathol.* 38: 45–51.
34. **Pauley, G. B.** 1969. A critical review of neoplasia and tumor-like lesions in mollusks. *Natl. Cancer Inst. Monogr.* 31: 509–539; **Dawe, C. J. and J. C. Harshbarger (Ed.).** 1969. *Neoplasms and Related Disorders of Invertebrate and Lower Vertebrate Animals*. National Cancer Institute Monograph 31, 772 pp.; **Farley, C. A.** 1976. Proliferative disorders in bivalve mollusks. *Mar. Fish. Rev.* 38: 30–33; **Mix, M. C., S. R. Trenholm, and S. I. King.** 1979. Benzo[a]pyrene body burdens and the prevalence of proliferative disorders in mussels *(Mytilus edulis)* in Oregon, pp. 52–64. in *Animals as Monitors of Environmental Pollutants*. National Academy of Science, Washington, D.C.; **Christensen, D. J., C. A. Farley, and F. G. Kern.** 1974. Epizootic neoplasms in the clam *Macoma balthica* (L.) from Chesapeake Bay. *J. Natl. Cancer Inst.* 52: 1739–1749.
35. **Farley, C. A.** 1969. Sarcomatoid proliferative disease in a wild population of blue mussels *(Mytilus edulis). J. Natl. Cancer Inst.* 43: 509–516; **Mix, M.** 1975. Proliferative characteristics of atypical cells in native oysters *(Ostrea lurida)* from Yaquina Bay, Oregon. *J. Invertebr. Pathol.* 26: 289–298; **Mix, M.** 1976. A review of the cellular proliferative disorders of oysters *(Ostrea lurida)* from Yaquina Bay, Oregon. *Prog. Exp. Tumor Res.* 20: 275–282; **Mix, M. C., H. J. Pribble, R. T. Riley, and S. P. Tomasovic.** 1977. Neoplastic disease in bivalve mollusks from Oregon estuaries with emphasis on research on proliferative disorders in Yaquina Bay oysters. *Ann. N.Y. Acad. Sci.* 298: 356–373.
36. **Lowe, D. M. and M. N. Moore.** 1979. The cytology and occurrence of granulocytomas in mussels. *Mar. Pollut. Bull.* 10: 137–141.

5 Ocean Pollution and Human Diseases

****

CHOLERA IN THE WESTERN HEMISPHERE — 1991

It was fiesta time, early January 1991, in a small coastal town on the outskirts of Lima, Peru, with a fleeting cornucopia of food, drink, and joy. One special dish that is a clear favorite among the inhabitants during any fiesta or special occasion is "ceviche" — prepared by marinating raw fish and shellfish for a few hours. This traditional dish gained its popularity long ago, in an earlier less complex time when people were not so numerous and coastal waters not so polluted. But, unfortunately, the risks of human disease from contaminated raw seafood have increased in proportion to the population size, and the ceviche served in that Peruvian village acted as a tiny but critical nucleus for catastrophic events that were to have effects far beyond the town.

As background information, it is important to note that an Asian freighter had anchored in the harbor during the previous week, and that the ceviche for the fiesta was prepared from raw shellfish harvested from that harbor. On the day following the celebration, many residents of the town became very ill, with vomiting, acute diarrhea, and extreme dehydration. Thirty-seven people died. A virulent Asiatic strain of a bacterial pathogen, Vibrio cholerae, was isolated from those stricken. The disease was diagnosed as cholera, a scourge from the dark ages that has never really disappeared from much of the world, but one that prospers where and when sanitary conditions are absolutely abominable.

The pathogen spread quickly to other nearby towns, then to Lima itself, where thousands became ill and hundreds died. Water supplies in the poorer districts became contaminated with the pathogen, and abysmal sanitary conditions added momentum to the expansion of the disease. Within weeks, sporadic outbreaks were occurring in other coastal areas of Peru, and travelers soon carried the pathogen to other South American and Central American countries. Within three months the disease was pressing against the Mexican border of the United States. By the end of 1991, more than 3,000 Peruvians had died from cholera, as had over 1,000 from other countries in South and Central America. In the summer of 1992, the epidemic showed signs of lessened intensity, although 62,000 new cases were diagnosed in the Americas in the first three months of 1992. By February 1993, Brazil had become a focus of the disease, with 32,313 cases and 389 deaths reported. By December 1993, cases of cholera in Latin America and the Caribbean had reached 700,000, with an estimated 6,400 deaths.

The United States, with reasonable levels of sanitation, has been spared most of the anguish and death caused by the disease. Less than 100 of its citizens acquired cholera, either during visits to South America or by eating contaminated food transported home by travelers. Of this total, 65 were infected and one died after eating contaminated seafood salad served on a plane bound from Lima to Los Angeles.

The likelihood of a major cholera outbreak in the United States is considered to be slight, since the disease is associated with primitive hygienic conditions not often found in this country. One exception might be among residents of squalid slums along the border with Mexico — areas that lack public water or sewage disposal systems.

From "Field Notes of a Pollution Watcher"
(C. J. Sindermann, 1994)

****

INTRODUCTION

Examples given in Chapters 3 and 4 of this section of the book offered evidence that some diseases and abnormalities of fish and shellfish seem to be associated with the intensity of contamination of the waters in which they live. A logical extension of any discussion of pollution and disease has to be aimed at the relationship of coastal/estuarine pollution and *human disease*. This is obviously where the average person's interest focuses.

Human illnesses can be acquired from contaminated seafood through two principal routes:

1. By eating raw or improperly processed fish and shellfish that have ingested and accumulated (or have had their flesh or external surfaces contaminated by) *micro-organisms* infective to humans. Included here would be microbial pathogens that cause several types of gastroenteritis, hepatitis, typhoid fever, and cholera.
2. By eating fish and shellfish that have been *chemically* contaminated by paralytic and other shellfish toxins derived from algae or by toxic levels of industrial chemicals — heavy metals and chlorinated hydrocarbons in particular.

Since the possibility of acquiring diseases from eating contaminated seafood has to be of concern to most of us, this chapter contains reasonably detailed examinations of the microbial and chemical contaminants of greatest potential harm to unwary, unwise, or deliberately adventuresome humans.

HUMAN MICROBIAL DISEASES TRANSMITTED BY CONTAMINATED SEAFOOD

As we dump more and more untreated or inadequately treated domestic sewage into rivers, estuaries, and coastal waters, the populations in those waters of microorganisms of human origin — bacteria and viruses in particular — will be increased. Dilution occurs as a result of river outflow, tidal flushing, and inshore currents, but may not take place fast enough to remove the risk of infection soon enough. Many bacteria that cause human disease neither reproduce nor survive very long in more saline ocean waters. However, they may not be killed instantaneously and so can constitute a threat to human health. Of particular concern are the microorganisms that cause cholera, typhoid, dysentery, skin infections, hepatitis, botulism, and eye and ear infections. Disease-causing bacteria of human origin, present in domestic sewage, may persist for days, weeks, or months in the intestines of fish, on the body surfaces or gills of fish and shellfish, and within the digestive tracts of shellfish, or on their gills, as well as in bottom sediments. Swimmers, skin divers, and fishermen obviously expose themselves to infection by venturing too close to ocean outfalls, sludge dumpsites, or badly degraded estuarine waters. Frequently, though, pollutants may be carried for miles by currents, so that it is difficult to determine which waters are safe and which are not, except by more or less continuous monitoring.

An added element of danger results from handling or eating uncooked fish and shellfish from polluted areas. Marine animals can and do ingest contaminated material, and certain shellfish may accumulate viruses and bacteria. Public health problems related to microbial contamination can be a major deterrent to full utilization of coastal resource species. Diseases such as typhoid and hepatitis have been transmitted by ingestion of raw shellfish from polluted waters[1]; hepatitis is an especially persistent problem.

Viral Diseases of Humans Transmitted by Shellfish

A number of epidemiological studies beginning in the 1950s have indicated a causal relationship between viral hepatitis and consumption of raw, fecally contaminated molluscan shellfish.[2] However, of the total number of cases of infectious hepatitis reported annually from all causes, the percentage transmitted by consumption of raw contaminated shellfish is a small, persistent, and controllable segment.[3] Despite the availability of information about disease risks, each year brings new reports of hepatitis outbreaks that can be traced to consumption of raw shellfish. As an early example, outbreaks of hepatitis A affecting almost 300 people, traced to eating raw oysters, occurred in Texas and Georgia in 1973.[4] The oysters were from Louisiana growing areas approved for harvesting under guidelines of the National Shellfish Sanitation Program. The source of contamination seemed to be floodwaters polluted several months earlier.[5] During the period 1961 to 1990, some 1,400 cases of oyster- and clam-associated hepatitis A were documented in the United States.[6]

Until 1974, all outbreaks of hepatitis associated with raw shellfish were thought to be caused by hepatitis A virus. In that year, hepatitis B virus was reported in repeated samples of clams, *Mya arenaria,* from one location on the Maine coast.[7] The site (one of 24 closed shellfish areas sampled) received untreated sewage from a coastal hospital in which two individuals with type B hepatitis were patients during the three months preceding the study. Transmission of the pathogen to previously unexposed clams in closed aquaria was achieved experimentally. The investigators concluded that clams must be considered potential vectors for hepatitis, and that under special circumstances they could serve as reservoirs for type B hepatitis virus as well as type A.

Viruses have been found experimentally to have variable, but in some instances surprisingly long survival time in saline waters — often under what would appear to be adverse conditions.[8] Rates of inactivation of enteric viruses in seawater increase with increasing temperature. For example, in one study,[9] 90% of poliovirus 2 was inactivated in sterile seawater in 48 days at 4°C, whereas 99.9% was inactivated in 30 to 40 days at 22°C. The virus survived four times longer in filter-sterilized seawater than in natural seawater, indicating that microorganisms in seawater (or their metabolites) are factors responsible for inactivation of the viruses. Important survival factors for viruses in seawater seem to be aggregation and adsorption onto particulates.

There is much research interest in procedures to inactivate or remove viruses from sewage treatment wastewater and sludge. Methods are mechanical, biological, and chemical, but none seems to be completely effective, and the number of complicating factors (for example, temperature, pH, particle size, electrical charge, flocculation, organic content) is daunting.[10]

Viruses affecting humans, then, constitute a critical problem for fishing or aquaculture operations in coastal/estuarine areas where even marginal domestic pollution exists — and because of nonpoint source runoff this includes most of the areas now used or planned for use in marine aquaculture. Additionally, viral contamination is and will be an important issue where treated sludges or other fecal degradation products are used for enrichment of growing areas, until large-scale, inexpensive techniques are available that will ensure total viral destruction. Shellfish purification (depuration) procedures must also take viral survival into account.

Bacterial Diseases of Humans Transmitted by Fish and Shellfish

Although viruses constitute a definite public health problem in utilizing inshore species as food, pathogenic bacteria also form a continuing threat, when raw or partially processed products are consumed by humans. Much attention has been paid during the past 40 years to the role of the vibrios, *Vibrio parahaemolyticus* and *Vibrio cholerae,* in outbreaks of gastroenteritis and cholera, respectively, that have been associated with consumption of raw or improperly processed seafood. While the vibrios are normal constituents of the inshore flora, their abundance may be increased facultatively by organic enrichment of coastal and estuarine areas, marine animals may carry or be infected by members of the genus, and seafood may be contaminated by improper handling. Most marine bacteria are not harmful to humans, but some of the vibrios can cause acute digestive disturbances, particularly when fish and shellfish carrying those bacteria are consumed raw or undercooked, and one species in particular, *Vibrio vulnificus,* can also cause fatal wound infections (Figure 24).

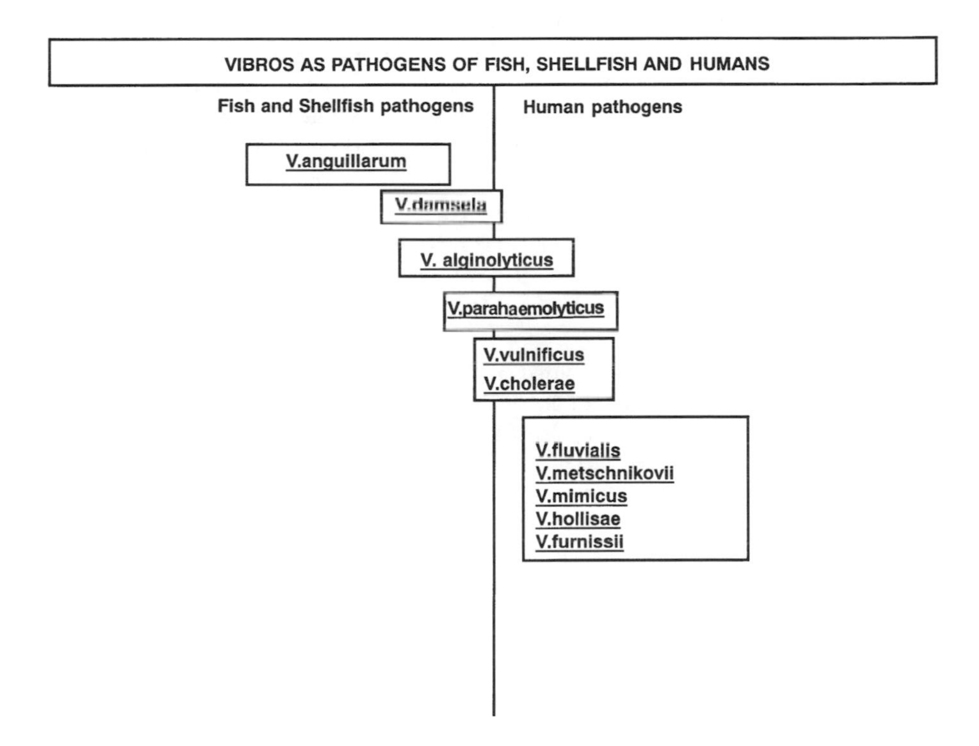

FIGURE 24. *Vibrios* as pathogens of fish, shellfish, and humans.

Beginning in the 1950s, summer bacterial gastroenteritis outbreaks in Japan have been traced to human ingestion of raw marine fish and invertebrates.[11] The largest outbreak, affecting 20,000 people, occurred in Niigata Prefecture in 1955 and was traced to eating cuttlefish from the Sea of Japan. Examples of the involvement of marine products in gastroenteritis outbreaks can be seen often in the statistics of the Japanese Ministry of Health and Welfare. The causative organism in many outbreaks was identified as the halophilic bacterium *Vibrio parahaemolyticus*. Numerous pathogenic and non-pathogenic strains have been isolated from coastal seawater, plankton organisms, bottom mud, and the body surfaces and intestines of marine fish and shellfish. Many strains have been recognized, and an extensive body of Japanese literature on *V. parahaemolyticus* has accumulated.

The Oriental custom of eating raw fish and shellfish (i.e., sushi and sashimi) has undoubtedly contributed to the severity of the vibrio problem there; 70% of all reported gastroenteritis outbreaks have been associated with *V. parahaemolyticus*. The organism was first recognized in Japan in 1951 as the cause of "shirasu food poisoning".[12] During the 20 years following recognition of the problem (1951 to 1971), over 1,200 technical papers on *V. parahaemolyticus* as well as several books were published. The natural habitat of the organism seems to be in estuaries rather than in the open sea. The infective dose for humans is 1 million to 1 billion organisms. *Vibrio parahaemolyticus* has a short generation time (9 to 11 minutes) — twice as fast as the common fecal bacterium *Escherichia coli* (at about 20 minutes) — which means that infective dose levels can be reached from an original population of only *10 organisms* in three to four hours — a remarkably short time.

Outbreaks of gastroenteritis traced to *V. parahaemolyticus* have been reported from many places in the world. To illustrate the nature of the problem, a random sampling made in 1976 of just a few of literally hundreds of outbreaks during the early 1970s included these events:

1. Two separate outbreaks occurred in 1974 on cruise ships in the Caribbean. One, which affected one-third of the ship's passengers, was traced to seafood cocktails, possibly contaminated by raw sea water from the ship's circulating salt-water system. The second outbreak, which involved over half of the passengers on another ship, was again traced to seafood that may have been contaminated by raw seawater during food preparation.[13]
2. An outbreak of *V. parahaemolyticus* gastroenteritis occurred in Panama in 1975, involving 30 people who had eaten a seafood dish prepared from fresh shrimp.[14]
3. Lightly cooked cockles were implicated in an outbreak in Singapore in 1975. Four strains of *V. parahaemolyticus* were isolated from patients, and *V. parahaemolyticus* was isolated from the cockles.[15]
4. Crabmeat cocktail served on a flight from Bangkok to London in 1972 sickened passengers and crew. *Vibrio parahaemolyticus* was suspected and was isolated from samples of raw crab claws obtained at the source, as well as from the crabmeat served on the flight.[16] (The reports did not reveal who piloted the plane during the latter part of the trip.)

5. Two outbreaks of gastroenteritis caused by *V. parahaemolyticus* occurred in 1971 in Maryland following picnics featuring steamed blue crabs. Investigation revealed that the cooked crabs had been delivered by truck, with baskets of *live* crabs loaded (unaccountably) on top of them, thereby allowing bacterial contamination of the cooked animals.[17]

These specific examples of seafood-associated illnesses are now ancient history — episodes that happened 20 years ago — but outbreaks of gastrointestinal disease that are believed to result from contaminated food still occur on modern cruise ships. According to the federal Centers for Disease Control (CDC) in Atlanta, Georgia, at least 830 people were made ill and one died as a result of four such shipboard epidemics in 1994 on cruise vessels sailing from South Florida ports. These recent outbreaks prompted the imposition of stricter CDC inspection guidelines late in that year.

An important observation that emerged from investigations conducted during the earlier period of the 1970s is that *V. parahaemolyticus* could cause outbreaks even when fish and shellfish were cooked. Improper processing procedures — undercooking, use of raw seawater to wash work surfaces, allowing raw seafood to drain onto cooked products, or placing cooked seafood on surfaces where raw marine animals have been shucked, cleaned, or sliced — can lead to ingestion of the pathogens by humans, with resultant gastrointestinal infections.

In addition to gastrointestinal disturbances, there have been disturbing reports of injury-induced tissue infections caused by marine vibrios, including *V. parahaemolyticus*. Case histories of such marine vibrio-related infections — some of them fatal and some requiring amputation — have been described in the literature before 1980,[18] and other lesser cases, in which *V. parahaemolyticus* was isolated from infected wounds, have also been reported.[19] Questions arose as to whether *V. parahaemolyticus* isolated from localized tissue infections acquired from coastal/estuarine waters were enteric pathogens with an altered route of entry, or whether they were "nonpathogenic" vibrios with previously unsuspected virulence. One extensive study indicated that isolates from wound infections were clearly similar to enteric forms isolated from cases of gastrointestinal illnesses in Japan and were unlike isolates from estuarine waters.[20] (A more recent question about these earlier reports of wound infections is whether the pathogens were actually *V. parahaemolyticus* or members of a species unrecognized before 1976, *Vibrio vulnificus,* to be discussed later.)

The early 1970s saw the peak in research and publication on *V. parahaemolyticus* by marine microbiologists; their efforts were diverted in the late 1970s and the early 1980s to a surge of research activity with *Vibrio cholerae* from coastal/estuarine sources.[21] Microorganisms with characteristics of *V. cholerae* were isolated from many estuaries in many countries. Extensive studies by the noted marine microbiologist Rita Colwell and her associates led to the conclusion that *V. cholerae* is a normal component of the flora of brackish waters, estuaries, and salt marshes of the temperate zone.[22] Other conclusions were that *V. cholerae* can occur in the absence of fecal contamination and that outbreaks can be expected in humans when proper food-handling techniques are not used. Sporadic outbreaks have occurred in a number of temperate zone countries — in Italy in 1973 and 1980, in Portugal in 1974, and in the U.S. (Louisiana) in 1978 (this was the first reported outbreak in the U.S. since 1911). Contaminated shellfish were implicated in each outbreak — mussels in Italy, cockles in Portugal, and crabs in Louisiana. Whereas *V. cholerae* may be a normal part of the brackish water microflora, its potential for causing human disease seems to be enhanced in heavily polluted shellfish-growing areas, especially if raw or improperly processed products are consumed or if confirmed cases of cholera have been reported in the adjacent towns.

As seen with certain other pathogens, whenever even one case of cholera occurs in a local human population, the danger of shellfish contamination will exist in surrounding waters. As an example, a study in Portugal[23] disclosed the presence of *V. cholerae* in 38% of 166 samples of molluscan shellfish taken in 1974 from the vicinity of Tavira, where a single case of cholera had been reported. This report was a sequel to an earlier paper[24] pointing out extensive pollution of shellfish-growing areas on the southern (Algarve) coast of Portugal. The *New York Times* (November 2, 1975) reported over 200 cases of cholera with 3 deaths in Coimbra, Portugal. Health authorities attributed the outbreaks to contaminated cockles from the Mondego River estuary.

The recent history of cholera in the U.S., just referred to, needs to be placed in the perspective of a longer time span. During much of the twentieth century, cholera was not reported in the U.S. — until 1973, when a case was reported from Texas. Since then sporadic small outbreaks have occurred: 11 cases in Louisiana in 1978, 2 cases in Texas in 1981, 17 cases on a Texas oil rig in 1982, and 13 cases in Louisiana and Florida in 1986. Most of the cases were associated with eating contaminated shellfish.

The most recent major outbreak of cholera in the Western Hemisphere began in 1991 in the port city of Chimbote, Peru. The first case, caused by a virulent Asian strain of the vibrio, was diagnosed in January. Early spread of the disease was attributed to eating fecally contaminated uncooked fish and shellfish in a popular dish called ceviche. Further spread was aided by ingestion of fecally contaminated drinking water as well as food, including raw vegetables. A little over a year later (March 1992), more than 3,000 Peruvians had died from the disease, and the epidemic had spread and continued to spread erratically through much of Central and South America. In early 1993, Brazil had become one of the foci of the epidemic, with 32,313 cases and 389 deaths reported, principally along the Atlantic coast of that country. By the end of that year, the grand total of cholera cases in Latin America and the Caribbean had reached 700,000, with an estimated 6,400 deaths.

Cases have been reported in Mexican cities near the U.S. border, and isolates of a *V. cholerae* strain identical to that found in Peru were recovered from oyster reefs in Alabama as early as September 1991, resulting in closure of the beds. The source of the pathogens was not determined, but human carriers from South America were suspected. In another incident, 65 (of 336) passengers on an Argentine airplane bound for Los Angeles were stricken with cholera in February 1992. One person died, and the outbreak was blamed (arguably) on eating contaminated seafood salad brought on board during a stop in Lima, Peru. Other isolated cases in the United States (totaling 24) have been associated with the South American epidemic — mostly travelers who ate contaminated seafood while in Central or South America, or family members who ate contaminated seafood transported home by the travelers.

The likelihood of a major cholera outbreak in the United States is considered to be slight, since the disease is associated with primitive hygienic conditions not often found in this country. One exception might be among inhabitants of poorer districts along the Mexican border, who lack public water or sewage disposal systems.

So much then for the anguish and death caused by the most notorious of the vibrios, *V. cholerae*.

The most recent *Vibrio* on the scene is one that can be severely pathogenic to some humans — *V. vulnificus*. The species was first recognized as a human pathogen in 1976, and the taxonomic group now includes some organisms formerly identified as *V. parahaemolyticus*.[25] Primary septicemias caused by *V. vulnificus* may result from ingestion of contaminated raw oysters or clams; wound infections with *V. vulnificus* result from contact with marine animals (lacerations from barnacles, shark bites) or with seawater containing the pathogens. Septicemias can cause death (61% of patients), especially among individuals with preexisting liver damage or immunodeficiencies, in a matter of hours or days.[26] Wound infections have a lower death rate (22%), but sometimes require amputations.

Two subgroups of the species *V. vulnificus* have been described: biogroup 1, associated with human infections, and biogroup 2, which includes strains pathogenic for eels. Biogroup 1 has been isolated from shellfish and seawater.[27]

It is important to emphasize that vibrios, including *V. parahaemolyticus, V. cholerae,* and *V. vulnificus,* are present in and on shellfish and in seawater, not as contaminants, but *as part of the normal microflora*. The *abundance* of these organisms, however, may be enhanced by organic loading of coastal/estuarine waters from human sources, or by augmentation, via sewage contamination, with pathogens from infected individuals.

Because some misguided humans (fortunately in diminishing numbers) persist in eating raw bivalve molluscs — especially oysters — outbreaks of seafood-borne gastrointestinal disease are grimly summarized every year in the aptly named "Morbidity and Mortality Report" of the federal Centers for Disease Control in Atlanta. The most recent epidemics (picked up from a CDC report by the news services in late January, 1995, as this section of the book was being edited) were of acute gastroenteritis in more than 100 people who ate sewage-contaminated raw oysters from Apalachicola Bay, Florida, and Galveston Bay, Texas, during and after the Christmas holiday period.

Other Microbial Diseases of Humans That May Have Some Association With Marine Pollution

Other bacterial genera, such as *Clostridium, Salmonella,* and *Shigella,* that are more directly pollution-related, should not be ignored in this discussion, since a single outbreak of disease related to any marine species can have a drastic impact on markets for *all* marine products.

Most studies of the relationship of fish to *Salmonella* infections in humans conclude that fish can serve as passive vectors of waterborne pathogens, and that the bacteria disappear from body surfaces and gut when the fish leave contaminated areas. An investigation in 1970 found that *Salmonella paratyphi* A survived for two weeks in filtered sterilized estuarine water from Chesapeake Bay and for two months in filtered, sterilized seawater from the Delaware coast.[28] Such a survival time, whether in sediments, in the water column, or in or on fish, could provide a passive mechanism for possible infection of humans, even without active infection of the fish.

Experimental infections with *Salmonella typhimurium* were obtained in mullet, *Mugil cephalus,* and pompano, *Trachinotus carolinus,* by two-hour exposure to 10^7 cells/ml in static aquarium systems.[29] Infections, in the form of hemorrhagic areas of the intestine from which pure cultures of *S. typhimurium* were recovered, were seen 10 to 14 days after exposure in some of the experimental fish. The organism was recovered from the alimentary tracts of the two fish species up to 30 days after exposure. It may be important to note, though, that the original isolates of *S. typhimurium,* on which the experimental exposures were based, were from the digestive tract of a mullet and not from an active mammalian infection or type culture collection. These results with *Salmonella* species indicate that fish can harbor the pathogens for appreciable periods following exposure, and that at least some exposed animals may actually become infected.

A recent government publication summarizing information on seafood-poisoning microorganisms listed nine bacterial genera as having been isolated from raw or processed seafood, and in some instances having caused human disease.[30] Present as contaminants in raw fish and shellfish, or as contaminants introduced during processing, were representatives of the genera *Vibrio, Salmonella, Shigella, Staphylococcus, Clostridium, Yersinia, Listeria, Campylobacter,* and *Escherichia (E. coli).* Of these, the vibrios are clearly the most significant from a public health perspective, with three species — *Vibrio parahaemolyticus, V. cholerae,* and *V. vulnificus* — definitely implicated as human pathogens, acquired from consumption of raw or inadequately cooked seafood. The other genera are contaminants introduced during processing and can be acquired from other kinds of animal products as well as from seafood. (As this chapter is being edited (January, 1995), accounts are appearing in national newspapers of an outbreak of *Shigella flexneri* on a Pacific cruise ship, causing illness in 640 people, with one death. The route of infection has not yet been determined, but inadequately refrigerated food, infected food handlers, or polluted seawater used to wash food preparation surfaces may be involved.)

CHEMICAL CONTAMINANTS TOXIC TO HUMANS

****

THE "STRANGE DISEASE" OF MINAMATA

Fortunately for the human species, there are still only a few known episodes of mass deaths and disabilities caused by industrial pollution of coastal/estuarine waters. Of those, the one that has received most attention occurred in and near the city of Minamata in southern Japan, with cases reported as early as 1953. Stricken individuals exhibited a variety of neurological disorders — trembling, numbness, paralysis, loss of control of body functions, and (in 40% of cases) death. Even more horrifying were the increasing numbers of abnormal children being born in the affected area, many retarded, blind, with misshapen limbs. Labeled the "strange disease" first, and later "Minamata Disease," the search for a cause was painfully slow, and was impeded by government–industry foot-dragging and denials that a problem existed.

Mercury contamination of Minamata Bay and its fish and shellfish populations by the effluents of a chemical production company was suspected as the cause and was reported as such in the scientific literature in 1959. It was not until 1968, however, that the Japanese government admitted officially that mercury contamination of fish and shellfish was the cause of the disease in human consumers and that the chemical company (a part of Chisso Corporation) was the source. Later investigations disclosed that during the period 1932 to 1965 Chisso had dumped 80 tons of mercury into the bay. The company denied legal responsibility until 1973, although beginning in 1959 it had begun offering poisoning victims and their families minuscule compensations (so-called "solatiums," which acknowledged corporate concerns but were not considered admissions of guilt). For all the suffering, disfigurement, and death that this company had caused, it paid "solatiums" of only $800 for a death, $280 per year for adult victims, and $83 for afflicted children! (These amounts were increased in the 1970s after public outrage and legal decisions in civil suits forced action by Chisso.) By 1975, 793 victims were designated officially, but about 2700 other people living in small fishing villages in the polluted zone surrounding Minamata City had symptoms of mercury poisoning, and an additional 10,000 residents were considered latent victims.

The Japanese mercury poisoning story was not confined to Minamata. In 1964, another outbreak of the "strange disease" due to mercury poisoning was discovered on the west coast of Japan near the city of Niigata. The source of contamination was traced to a factory of the Showa Denko Corporation that was using the same acetaldehyde production process as that used by Chisso in Minamata — with mercury being dumped into adjacent river waters. Cases were mostly confined to fishermen and their families, for whom fish was a staple in their diets. Fortunately, the scale of contamination was less than that of Minamata Bay, with fewer deaths and disabilities. Then in 1973 a third outbreak of the disease, involving only ten cases, was reported from the coastal area bordering the Ariake Sea, 40 km north of Minamata Bay — an announcement that touched off near panic in all of Japan.

Events during the course of these gross and culpable polluting episodes seem to follow an almost predictable course. The polluting company, deliberately or through negligence, dumps toxic wastes into rivers or estuaries. Fish and shellfish may be affected, in that abnormalities appear and populations may decline. Fish eating birds and mammals (including humans) may be affected by ingested toxic chemicals, so that sickness and death may result. Medical scientists address the problem after growing public concern results in release of research funds. Investigations focus on suspected chemicals, and evidence accumulates about a specific contaminant and its likely source.

The polluting company during this period stoutly denies that its effluents can maim or kill, even when it has evidence from its own captive scientists that the company's effluents are responsible (as did the Chisso Corporation early in the Minamata incident). Governments and their regulatory agencies during this phase of denial are reluctant to act against the industry in the absence of truly overwhelming data and vigorous public outcry against the miscreants. Civil suits instituted by individuals or groups may eventually resolve the impasse, if decided in favor of the victims, and if endless appeals by the polluting industry are denied. The proffered settlements are usually far too small to compensate for damages; the polluting company then pleads poverty and begs for a government bailout; the final stage is threatened or actual declaration of bankruptcy after the company has spun off the more profitable of its subsidiaries as new companies.

The sequence of events in these major pollution episodes, often encompassing decades, seems to follow this common pattern — almost a formula — augmented occasionally with variations designed to avoid admission of responsibility for environmental degradation, with accompanying damage to resources and people. One of the few positive aspects of this prolonged struggle of victims against polluting industries has been the reinforcement of an important legal concept that had been developed during repeated earlier confrontations with polluters of

Japanese waters, which states: "In the absence of conflicting clinical or pathological evidence, epidemiological proof of causation [of human disease] suffices as legal proof of causation." This hard-won concept will be important in any future major contamination episode anywhere in the world that results in human illness.

From "Field Notes of a Pollution Watcher"
(C. J. Sindermann, 1983)

****

Polluted estuarine and marine coastal waters may cause public health problems of a chemical nature through three principal routes:

1. Ingestion of the chemically contaminated water directly
2. Direct physical contact with toxic or noxious chemicals in the water or in spray
3. Consumption of fish and shellfish that contain toxic levels of harmful chemicals

Our consideration here will be limited to category 3, since this is the resource-related aspect of the problem.

Industrial Chemical Contaminants

Fish and shellfish may accumulate dangerously high levels of pesticides, heavy metals, and other potentially toxic chemicals in grossly polluted waters. Of the chemicals that could occur in seafood at levels harmful to humans, mercury has justly received the most attention. The horrors of Minamata Disease, caused by mercury contamination of fish and cultivated shellfish in a bay in southern Japan, were publicized over 30 years ago on television, in news magazines, and in several books.[31] Severe permanent neural damage characterized those individuals most seriously afflicted. Partly as a consequence of this mass poisoning, increased surveillance of mercury and other heavy metals in all kinds of seafood lessens the likelihood of another Minamata incident, although whenever marine products are grown near industrial operations there is always a risk of chemical contamination, through negligence, deliberate dumping, or accidental spills. It is not feasible to provide adequate continuous chemical surveillance of every localized area where fish and shellfish are produced — especially since some of the analytical methods are very time-consuming and costly, and the toxic action levels for some contaminants are not fully understood.

Although mercury has achieved the most notoriety among heavy metal contaminants of food, public health problems have also been created by toxic levels of other metals. Ingestion of cadmium-contaminated water resulted in a disease called "itai-itai" in Japan during the decade following World War

II.[32] The problem developed from the use of cadmium-contaminated river water in two towns near metal mines; the contaminant caused severe disturbance of calcium metabolism, characterized by neurologic symptoms and extreme skeletal fragility.

A detailed evaluation of potentially harmful metals in fish and other seafood was made recently by an international joint "Group of Experts on the Scientific Aspects of Marine Pollution" (GESAMP). This prestigious U.N.-sponsored group reached a number of major conclusions about an array of metals:[33]

Mercury: "Populations with high fish intake or intake of fish with a high methylmercury content can easily exceed the World Health Organization/Food and Agriculture Organization (WHO/FAO) provisional tolerable intake level. Pregnant women constitute a special risk group."

Cadmium: "Only under exceptional circumstances will cadmium intake from fish constitute an important part of the total daily intake via food. High consumption of certain shellfish may increase considerably the intake, and, over many years, may increase cadmium concentrations in the kidney to toxic levels."

Arsenic: "Exposure to arsenic via seafood may be substantial. Most of this arsenic is in the form of arsenobetaine, which is considered relatively nontoxic. Extreme seafood consumption may give rise to an intake of several hundred micrograms of inorganic arsenic per day, an exposure level which over a lifetime may be related to a significant increase in skin cancer."

Lead: "Lead in seafood does not greatly contribute to the daily intake of lead, but other sources of lead (such as paint) will be additive."

Tin: "The contribution of seafood to the daily intake of tin is low. However, more data are needed for trimethyltin, which is synthesized by marine organisms and which may produce neural pathology in humans."

Selenium: "Selenium does not pose a toxicological problem, but its interaction with mercury compounds may be biochemically significant."

The GESAMP report on metals in seafood (as summarized) indicated greatest current concern with mercury and arsenic. Statements from the summary about the effects of mercury are:

> Mercury, in the form of methylmercury (MeHg), is still considered a prime pollutant in fish, including marine fish. Its possible implications for human health are important, and more and more emphasis is being put on the study of developmental effects, as observed in young children prenatally exposed to low concentrations of MeHg.
>
> Population groups consuming one normal fish meal/day (150 g fish) will reach the provisionally tolerable weekly intake (PTWI) of 200 μg mercury even when MeHg concentrations in the fish consumed are very low. For people who

eat only one seafood meal per week (about 20 g fish/day), the PTWI will not be exceeded, even when the average MeHg concentration (in the fish) is very high.

The effects of arsenic were summarized as follows:

Seafood is the predominant source of human arsenic intake. From the toxicological point of view there are two forms of arsenic in marine organisms which should be considered, namely arsenobetaine, which is the dominant form in most seafood, and inorganic arsenic, which constitutes 2 to 10% of the total arsenic content in seafood. Inorganic arsenic is by far the most toxic form and has given rise to skin lesions, such as hyperkeratosis, hyperpigmentation and skin cancer, peripheral blood vessel pathologies, effects on the central nervous system, and chromosome damage. In cases of extreme consumption of seafood, the intake of inorganic arsenic would reach levels at which the increased risk for skin cancer is definitely no longer negligible.

Another aspect of chemical pollution of seafood that has not been fully appreciated until recently is the possible long-term effect on humans of consumption of low levels of contaminants in food. Some contaminants are readily metabolized and excreted; others may accumulate in storage tissues as a result of continued ingestion. Certain of the heavy metals may accumulate, if the ingestion rate exceeds rates of detoxification and excretion. Several of the fat-soluble contaminants — especially the chlorinated hydrocarbons — can build up in humans as well as other animals, if the diet provides continuing low-level dosages.

Extensive studies have been made of the effects of chlorinated hydrocarbons — especially DDT and PCBs — on humans. In the preceding chapter, the effects on reproduction in fish resulting from contamination of adult individuals by chlorinated hydrocarbons were examined, with the finding of decreased viable hatch and increased occurrence of genetic/developmental abnormalities in the offspring. Contamination of human females by chronic dietary intake of PCB-contaminated fish can also result in deleterious effects on offspring. One of the most extensive studies of effects of human exposure to PCBs in food fish was reported recently from a major study in the Midwest.[34] The testing focused on mothers and newborn infants who had been chronically exposed to PCBs through consumption of fish from Lake Michigan. Maternal health effects included tendencies toward increases in anemia, edema, and susceptibility to infections. Effects on infants via transplacental transmission of PCBs and PCB contamination of maternal breast milk included delays in developmental maturation, decreased responses to stimuli, and decreased visual recognition memory. The startling estimation was made that "...if the input of PCBs into humans were to be stopped in a given maternal generation, the contaminants would still be transmitted in measurable concentrations...for at least five generations...."

****

EARLY EXPERIENCES WITH PCBS IN NEW ENGLAND

From high school to a production job in the Sprague Electric Company — that was a normal career progression for many 18-year-olds in the western Massachusetts mill town where I was born too many years ago. The company was one of the early users of PCBs, a versatile "oil" that filled the condensers and capacitors for a primitive pre-transistor electronic industry. I worked the night shift, sometimes up to my elbows in that magic fluid, filling those thousands of metal containers before riveting their terminals and sending them along for "degreasing" to a cohort standing over a vat of boiling solvent. It was like a scene from Hades, and some of my buddies didn't tolerate the oil too well. Skin rashes, headaches, dizziness, bronchitis, eye irritations and nausea drove some back to other less stressful departments, with lower pay scales. Outside the walls, residues of the oil were part of the effluent that emptied into the nearby Hoosac River, a terribly abused slimy tributary of a similarly degraded major river, the mighty Hudson. [Later it was discovered that the company had been also, quite legally at the time, burying drums of PCBs on the shores of the Hoosac.]

This was in post-neanderthal times — the very early 1940s — long before the first faint stirrings of any kind of environmental ethic. We knew nothing in those days about ecology or toxicology. PCBs were not recognized as a serious problem until the 1970s. Evidence of harmful effects on humans and other animals accumulated slowly, especially effects on reproductive processes.

Unknown to most of us working the night shift in that decaying factory building, a remarkably similar scenario was unfolding in the basement rooms of Sprague's arch-competitor — a foreign-based company located in New Bedford, with an improbable name, Cornell-Dubilier. The one significant and critical difference between the two companies was that in New Bedford the manufacturing wastes, including PCBs, were dumped directly into the harbor, where even now, decades later, commercial fishing is restricted and PCB levels in resource species are still greater than action levels.

I left that Massachusetts mill town forever soon after World War II, but I have often wondered about what happened to all of those innocent co-workers, male and female, whose life histories placed them at risk from long-term exposure to high environmental levels of PCBs. Did they have normal kids? What about their cancer histories? What have been their survival rates? That no-doubt fascinating story has never been told; the appropriate detailed research has not been done, and the consequences of those early and prolonged contacts with a toxic industrial chemical are rapidly receding from living memory.

From "Field Notes of a Pollution Watcher"
(C. J. Sindermann, 1990)

****

Possible toxic effects of chlorinated organic pesticides in foods have been discussed on many occasions. Conclusions from an earlier government study — made more than 20 years ago — were that "Despite the widespread occurrence of persistent pesticide residues in the world fauna, their magnitude is for the most part too small to have any known significant effects on human health."[35] (Emphasis here should obviously be on the word "known"). Even though indiscriminate use of persistent pesticides is coming under some measure of control in the United States, their use in other parts of the world is expanding, and contamination of the world's oceans is continuing and may even be increasing. Because of their persistence in the environment and their accumulation by successive levels of food chains, pesticides continue to be threats in near-shore ocean areas, including those devoted to marine aquaculture and those important as nursery areas for fish and shellfish. The sublethal effects of long-term exposure to low levels of pesticides in the diets of most marine animals are incompletely understood.

It might be well at this point to discuss the matter of carcinogens (cancer-causing agents) in the marine environment, since among the sublethal effects of chemical contaminants in estuarine and coastal waters are those that involve carcinogenic properties of the chemicals. Marine animals themselves may be affected, or, more significantly, the carcinogens may be accumulated in fish and shellfish that are then consumed by humans. The public health risks from ingestion of carcinogen-contaminated marine products can easily be appreciated (intuitively), but the extent of the present contamination of seafood is poorly documented, and the long-term effects of eating such contaminated products are unknown, except in the negative sense that no reported cases of human cancer have been traced directly thus far to ingestion of contaminated fish or shellfish.

Roughly 40 chemicals or groups of chemicals are considered to be carcinogenic for humans. Best known are arsenic and certain other metals, PCBs, DDT derivatives, benzo[*a*]pyrene (BaP) and other polycyclic aromatic hydrocarbons (PAHs), dioxin, and toxaphene. Some of them accumulate, occasionally at high levels, in fish and shellfish in areas of local pollution, or in larger enclosed bays and estuaries. Risk assessments with pollutants such as PCBs have suggested increased risk of cancer as a consequence of consumption of fish containing high levels of the contaminant,[36] but direct relationships between seafood consumption and cancer have not yet been demonstrated.[37]

There are, however, disquieting pieces of information that emphasize the importance of determining the levels and effects of chemical carcinogens from the contaminated marine environment. Some investigations have focused on the PAHs, particularly BaP, which is highly carcinogenic. In one study,[38] BaP levels as high as 121 ppm of dry sediment were found in the immediate vicinity of a Pacific coast sewage treatment plant, with diminishing concentrations at increasing distances from the outfall. In a related study, the authors examined BaP levels in mussels, *Mytilus edulis,* from stations near Vancouver, British Columbia, and found some values as high as 215 ppm wet weight of tissue.

Arsenic has been implicated in recent studies as a carcinogen. Studies in Sweden[39] indicated an increased risk of skin cancers (squamous carcinomas) associated with consumption of fish with high levels of inorganic arsenic. Higher incidences were found in fishermen, when compared to other occupational groups. An earlier paper[40] also reported frequent occurrences of skin cancers in deep-sea fishermen and fishing industry wharf workers. Other factors, such as exposure to ultraviolet (UV) radiation, may of course be involved in the genesis of skin tumors, so the conclusion must be that available epidemiological data do not support or refute an association between cancer and arsenic intake via fish.

Contaminant-related events with direct or potential public health significance seem to be increasing in frequency in coastal/estuarine waters. While some of this change may be due to greater public awareness and some improvement in surveillance, it is probable that the remarkable expansion in synthetic chemical production and use in the past three decades has contributed substantially. Synthetic chemicals that simply did not exist even a decade or two ago are now being viewed with some alarm as environmental contaminants. Many such chemicals have been dumped indiscriminately in rivers and estuaries and some are still being released. Disclosure of most pollution problems is accidental, and even after such disclosure, regulatory agencies frequently encounter strong resistance from the contaminating industries.

Public alarm about contaminants in food and water that may affect human health can lead to closure of industries or modifications of production methods, but usually only after clear evidence of danger to humans has appeared, and has been widely publicized in news media. There have been several incidents in North America during the past three decades that illustrate a common sequence of events: (1) release of toxic chemicals in the absence of surveillance and control, and with little knowledge of or concern about their public health effects; (2) some preliminary accidental or fortuitous indication of danger to humans or to resources; (3) vigorous denial of danger or responsibility by the polluting industry involved; (4) preliminary investigation and half-hearted action by regulatory agencies; (5) legal delaying tactics by the polluting industry; (6) reluctant compliance by the offending industry in the presence of mounting data, advisory legal opinions, and a rising crescendo of expressed public concern; and (7) grudging and usually minuscule payments (mostly absorbed by lawyers) to settle damage claims. Examples of this sequence include the release of PCBs in the Hudson River and Great Lakes, the release of Kepone in the James River in Virginia, and the release of phosphorous in Placentia Bay, Newfoundland.

The PCB story in the Hudson River is still unfolding. It includes deliberate long-term release of PCBs into the upper river by two units of the General

Electric Company; the finding of dangerously high levels of PCBs in fish; the closing of the river to fishing (except for certain anadromous species); the reluctant reduction of contamination and token cleanup efforts by the offending industry after much legal foot-dragging; and (the ultimate insult) the successful attempt by that industry to shift most of the financial burden of adequate cleanup to the U. S. taxpayers.

Kepone, a highly toxic and persistent insecticide, was deliberately discharged into the upper James River (an important oyster-producing area) over a period of 16 months in 1974–1975 by a subsidiary of Allied Chemical Company. Only after obvious toxic effects on chemical plant workers were disclosed was there any concern about pollution of waters and shellfish by the plant effluents. The producing facility was closed, and the river was also closed to fishing for an extended period.

The Placentia Bay (Newfoundland) phosphorous contamination event was first disclosed by observations of "red herring" and extensive herring mortalities in that bay in 1969. A new industrial operation had begun in 1968, and an investigation revealed that it had been releasing phosphorous into the bay (which is also a very important fishing area). Again, as with the Kepone event, the plant was closed temporarily, as was the fishery, in the presence of a clear danger to public health.

So, in summary, while there is little evidence of significant direct danger to human life in existing chemical contaminant levels in marine resource species (except for very localized incidents of gross pollution such as those just described), there are instances (such as mercury in black marlin, large halibut, and swordfish, and PCBs in striped bass) of contaminant levels high enough to warrant attention, further study, and possibly controlled consumption. There is also a *great* need for much more scientific examination of possible long-term sublethal effects on humans caused by contaminants in food. Sufficient evidence now exists about carcinogenic, mutagenic, and other long-term toxic effects of many industrial chemicals to warrant more attention and reasoned action, even in the absence of incontrovertible proof of risk to public health.

Natural Biotoxins

An additional chemical public health problem is that of biotoxins — paralytic shellfish poisoning (PSP) and ciguatera, in particular. Periodic toxicity to humans has long characterized molluscan species from certain coastal areas, and some species of fish from particular locations in tropical and subtropical waters have long been known to be toxic. However, recent changes in the distribution and intensity of fish and shellfish toxicity have taken place; these may be related to human modification of coastal and atoll waters. As examples, shellfish poisoning spread southward along the New England coast during the 1970s; mussel toxicity on the northeast coast of England increased during the same period.[41] Other European reports of PSP outbreaks seem to be related to high nutrient levels in coastal and harbor areas. Recently, too, a red tide on the west coast of Mexico caused by the dinoflagellate *Gymnodinium catenatum* resulted in an outbreak of PSP, with some mortalities in humans who consumed toxic oysters, *Crassostrea iridescens,* and clams, *Donax* sp. Previous algal blooms in that area had not been toxic.[42] Other reports of PSP outbreaks have come from many parts of the world, including both coasts of North America, western Europe, South America, Japan, South Africa, and New Zealand.

Whatever the explanation might be for the observed geographic changes in the distribution and abundance of toxin-producing microalgae (and it is increasingly difficult to rule out human intrusion completely), economic and public health problems have been created. Existing toxin-monitoring programs have had to be extended and new ones created to determine when closure of fisheries would be necessary to protect seafood consumers. Especially hard hit during recent toxic episodes have been the clam, scallop, and mussel industries of New England, the scallop aquaculture industry of northern Japan, and the mussel aquaculture industry of northwestern Spain. During periods of peak toxicities, these industries have come to a standstill, and a short-term "spillover" effect, based on public perceptions, has resulted in difficulties in marketing other seafood products.

Shellfish have been identified as passive carriers in a number of outbreaks of human illnesses due to natural toxins. The list begins with the well-known and widespread PSP, which is a consequence of blooms of toxic dinoflagellates of the genus *Protogonyaulax* (formerly *Gonyaulax*), and with "neurotoxic shellfish poisoning," caused by recurrent blooms of *Ptychodiscus brevis* (formerly *Gymnodinium breve*), which are also responsible for massive fish kills in the Gulf of Mexico.[43] Next in order of appearance is "diarrhetic shellfish poisoning," due to other genera of toxic dinoflagellates (*Dinophysis, Prorocentrum,* and others). This form of poisoning was first reported in Europe in 1961.[44] It became important in the late 1970s, after outbreaks in Japan and Europe, and persisted as a significant problem in the 1980s.[45] In December 1987, a "new" toxin, domoic acid, which causes "amnesic shellfish poisoning," was found in mussels from Prince Edward Island in the Gulf of St. Lawrence and was responsible for 129 cases of poisoning and two deaths in Canada.[46] Domoic acid can affect the brain and the nervous system of humans. It is concentrated, during blooms, in the digestive gland of shellfish, especially mussels, and (during the Canadian outbreak) its origin was reported to be a persistent bloom of the diatom *Nitzschia pungens* in mussel culture areas of Prince Edward Island.[47] The toxic event dealt a severe blow in 1988 to the emerging mussel aquaculture industry of the island.[48] Some key aspects of various forms of natural algal toxins are shown in Table 2.

As the reported occurrences of algal blooms and the associated phenomena of fish kills and shellfish toxicity increase, more and more research is funded,

TABLE 2
Types of Shellfish Poisoning

Illness	Symptoms	Cause	Description
Paralytic shellfish poisoning	Numbness of lips, face, and extremities; visual disturbance; staggering gait; difficulty breathing; paralysis	Species of the dinoflagellate *Protogonyaulax* (formerly *Gonyaulax)*	About 12 forms purine-derived saxitoxins; water soluble; acts by blocking sodium channel needed to transmit nerve impulses
Neurotoxic shellfish poisoning	Nonfatal but unpleasant neurotoxic toxic symptoms; strong action on cardiovascular system	*Ptychodiscus brevis* (formerly) *Gymnodinium breve)*	Toxins have unusual polycyclic ether skeletons
Diarrhetic shellfish poisoning	Cramps; severe diarrhea; nausea; vomiting; chills; death rare	Species of dinoflagellates, *Dinophysis* and *Prorocentrum* in particular	*Dinophysis* toxin and okadaic acid; large, fat-soluble polyethers; can move across cell membranes and make them "leak"
Amnesic shellfish poisoning (domoic acid)	Nausea; vomiting; muscle weakness; disorientation; loss of short-term memory	Reported to be diatom *Nitzschia pungens;* toxin also found in species of the red alga, *Chondria*	Nonessential amino acid; mimics glutamic acid; affects brain and nervous system; found in digestive glands of contaminated shellfish

Adapted from Pirquet, K. T., *Can. Aquacult.*, 4(2), 41, 1988.

and understanding accelerates. Several major international conferences on the subject have been held, and the results have been published as a continuing series of books.[49] Some recent findings are:

1. Dinoflagellate toxicity, which may be harmful to higher animals, has been found in at least 20 shellfish species representing 10 genera. Paralytic shellfish poisoning persists as a recurrent problem in some coastal waters of North America and Europe, and it occurs sporadically in other temperate and tropical areas.
2. The previously unrecognized type of shellfish poisoning (diarrhetic shellfish poisoning), with diarrhea as the dominant symptom, was first described in Europe in 1961 and was subsequently reported from several locations in Japan and in Europe. It must be noted, though, that shellfish-associated enteric disturbances have a long history, and this type of poisoning may have been present much earlier, but not diagnosed correctly.
3. Resting cysts of the PSP toxin-producing species *Protogonyaulax tamarensis* from bottom sediments were found to be ten times more toxic than were motile stages. Cysts may be ingested by shellfish, causing toxicity even when blooms are not apparent.
4. Markets for shellfish grown and harvested for export are especially sensitive to toxic outbreaks and may react violently. An example of this phenomenon was seen in 1976, when an epidemic of PSP was reported in Spain, France, Italy, Germany, and Switzerland — the consequence of consumption of mussels grown in northwestern Spain. Exports ceased for an extended period, and a monitoring system was instituted by Spain. A subsequent outbreak of diarrhetic shellfish poisoning in France affected 10,000 people in 1984 and 2,000 people in 1985.
5. Blooms of algae, known previously as causes of only sporadic or localized outbreaks, seem to be occurring with increasing frequency and intensity. An example would be blooms of the naked dinoflagellate *Gyrodinium aureolum,* observed to cause mortalities of marine organisms in England in 1978, in Scottish salmon cages in 1980, and in wild as well as captive fish on the coast of Norway in 1981 and in 1982. Another example would be mortalities of farmed salmon in Scotland in 1979 and in 1982 caused by blooms of a species of *Olisthodiscus* or *Chattonella*. These agents had been recognized earlier in connection with unusual blooms, but occurrences were rare and geographically restricted.

Monitoring and related research in a number of countries seem to indicate that episodes of toxicity in molluscan shellfish are increasing in geographic distribution, variety, and severity. Effects are especially critical when blooms occur in or near aquaculture operations. Organic enrichment of coastal/estuarine waters will certainly enhance algal blooms of all types and could contribute to the toxicity problem. Transport of toxic algae, either in ballast water or with introduced animals or plants, may establish new foci for eventual toxic blooms, and some evidence exists to support this phenomenon. Monitoring and surveillance systems have become a necessity as foreign markets have been developed, but the control of blooms (hence, toxicity) is not now and may never become a reality.

CONCLUSIONS

Public health matters are by definition important to all of us, so it would be logical in concluding this chapter on ocean pollution and human diseases to look briefly beyond present data, and even beyond present operational concepts, at the potential for future harmful effects of contamination of coastal living resources. We have in this chapter cited several instances, such as Minamata Disease in a coastal area of Japan and cholera in a coastal area of Portugal, of localized resource-related damage to the human population caused by pollution. These may be dramatic illustrations of insidious long-term damage that can only be speculated about now. We know, for example, that some heavy metals and many hydrocarbons can be carcinogenic and mutagenic, or can produce physiological and biochemical changes in test animals. Much of the information has been derived from acute exposures of laboratory animals to specific toxicants, but information about effects of long-term chronic exposures is already appreciable, and it suggests that continuous exposure to low levels of contaminants can be dangerous to human health. Of course, contamination of fishery products from estuaries and coastal waters is only a small part of the total problem of chemical and microbial contamination of food, but because coastal waters are the recipients of many chemicals of terrestrial origin, because marine organisms can selectively accumulate some contaminants (especially at higher trophic levels), and because seafood constitutes a significant part of the diet in a number of countries, the long-term effects of seafood contamination on human health cannot be ignored.

REFERENCES

1. **Mason, J. O. and W. R. McLean.** 1962. Infectious hepatitis traced to the consumption of raw oysters. An epidemiologic study. *Am. J. Hyg.* 75: 90.
2. **Ross, B.** 1956. Hepatitis epidemic transmitted by oysters. *Sven. Laekartidn.* 53: 989–1003; **Dougherty, W. J. and R. Altman.** 1962. Viral hepatitis in New Jersey. *Am. J. Med.* 32: 704–736; **Richards, G. P.** 1985. Outbreaks of shellfish-associated enteric virus illness in the United States: requisite for development of viral guidelines. *J. Food Prot.* 48: 815–823.
3. **Liu, O. C., H. R. Seraichekas, and B. L. Murphy.** 1966. Viral pollution and self-cleansing mechanism of hard clams, pp. 419–437. in Berg, G. (Ed.), *Transmission of Viruses by the Water Route*. Wiley-Interscience, New York.
4. **Hughes, J. M.** 1979. Epidemiology of shellfish poisoning in the United States, 1971–1977, pp. 23–28. in Taylor, D. L. and H. H. Seliger (Eds.), *Developments in Marine Biology*. Vol. 1. *Toxic Dinoflagellate Blooms*. Elsevier/North-Holland, New York.
5. **Portnoy, B. L., P. A. Mackowiak, C. T. Caraway, J. A. Walker, T. W. McKinley, and C. A. Klein.** 1975. Oyster-associated hepatitis: failure of shellfish certification programs to prevent outbreaks. *JAMA* 233: 1065–1068.
6. NOAA (National Oceanic and Atmospheric Administration). 1991. The 1990 National Shellfish Register of Classified Estuarine Waters. NOAA, Off. Oceanogr. Mar. Assess., Rockville, MD, 99 pp.; **Richards, G. P.** 1987. Shellfish-associated enteric virus illness in the United States, 1934–1984. *Estuaries* 10: 84–85.
7. **Mahoney, P., G. Fleischner, I. Millman, W. T. London, B. S. Blumberg, and I. M. Arias.** 1974. Australia antigen: detection and transmission in shellfish. *Science* 183: 80–81.

8. **Metcalf, T. F. and W. C. Stiles.** 1966. Survival of enteric viruses in estuary waters and shellfish, pp. 439–447. in Berg, G. (Ed.), *Transmission of Viruses by the Water Route*. Wiley-Interscience, New York.
9. **Gerba, C. P. and G. E. Schaiberger.** 1975a. Effect of particulates on virus survival in seawater. *J. Water Pollut. Control Fed.* 47: 93; **Gerba, C. P. and G. E. Schaiberger.** 1975b. Aggregation as a factor in loss of viral titer in seawater. *Water Res.* (G.B.) 9: 567; **Schaiberger, G. E., C. P. Gerba, and E. G. Estevez.** 1976. Survival of viruses in the marine environment, pp. 97–109. in Meyers, S.P. (Ed.), *Proceedings of the International Symposium on Marine Pollution Research,* Gulf Breeze, FL.
10. **Cooper, R. C.** 1975. Waste water management and infectious disease. II. Impact of waste water treatment. *J. Environ. Health* 37: 342–353; **Berg, G.** 1976. Microbiology — detection, occurrence, and removal of viruses. *J. Water Pollut. Control Fed.* 48: 1410–1416.
11. **Iida, H., T. Iwamoto, T. Karashimada, and M. Kumagai.** 1957. Studies on the pathogenesis of fish-borne food-poisoning in summer. II. Studies on cholinesterase inhibition by culture filtrates of various bacteria. *Jpn. J. Med. Sci. Biol.* 10: 177–185; **Aiso, K. and M. Matsuno.** 1961. The outbreaks of enteritis-type food poisoning due to fish in Japan and its causative bacteria. *Jpn. J. Microbiol.* 5: 337–364.
12. **Fujino, T., Y. Okuno, D. Nakada, A. Aoymama, K. Fukai, T. Mukai, and T. Ucho.** 1953. On the bacteriological examination of Shirasu-food poisoning. *Med. J. Osaka Univ.* 4: 299–304.
13. **Craun, G. F.** 1975. Microbiology — waterborne outbreaks. *J. Water Pollut. Control Fed.* 47: 1566–1569.
14. **Kourano, M. and M. A. Vasquez.** 1975. The first reported case from Panama of acute gastroenteritis caused by *Vibrio parahaemolyticus*. *Am. J. Trop. Med. Hyg.* 24: 638.
15. **Foo, C. K.** 1975. Source of *Vibrio parahaemolyticus* infection. *Singapore Med. J.* 15: 188 (Microbiol. Abstr. 10, B6866).
16. Anonymous. 1972. *Vibrio parahaemolyticus* gastroenteritis United Kingdom. U.S. Public Health Service, *Morbid. Mortal. Weekly Rep.* March 25, pp. 99, 104.
17. **Colwell, R. R., T. E. Lovelace, L. Wan, T. Kaneko, T. Staley, P. K. Chen, and H. Tubiash.** 1973. *Vibrio parahaemolyticus* — isolation, identification, classification and ecology. *J. Milk Food Technol.* 36: 202–213.
18. **Fernandez, C. R. and G. A. Pankey.** 1975. Tissue invasion by unnamed marine vibrios. *JAMA* 233: 1173–1175.
19. **Poores, J. M. and L. A. Fuchs.** 1975. Isolation of *Vibrio parahaemolyticus* from a knee wound. *Clin. Orthop.* 106: 245–252.
20. **Twedt, R. M., P. L. Spaulding, and H. E. Hall.** 1969. Morphological, cultural, biochemical, and serological comparison of Japanese strains of *Vibrio parahaemolyticus* with related cultures isolated in the United States. *J. Bacteriol.* 98: 511–518.
21. **DePaola, A.** 1981. *Vibrio cholerae* in marine foods and environmental waters: a literature review. *J. Food Sci.* 46: 66–70.
22. **Colwell, R. R., R. J. Seidler, J. Kaper, S. W. Joseph, S. Garges, H. Lockman, D. Maneval, H. Bradford, N. Roberts, E. Remmers, I. Huq, and A. Huq.** 1981. Occurrence of *Vibrio cholerae* serotype 01 in Maryland and Louisiana estuaries. *Appl. Environ. Microbiol.* 41: 555–558.

23. **Ferreira, P. S. and R. A. Cachola.** 1975. *Vibrio cholerae* El Tor in shellfish beds of the south coast of Portugal. *Int. Counc. Explor. Sea,* Doc. C.M. 1975/K:18, 7 pp.
24. **Cachola, R. and M. C. Nunes.** 1974. Quelques aspects de la pollution bactériologique des centres producteurs de mollusques de l'Algarve (1963–1972). *Bol. Inf., Inst. Biol. Marit.* 13: 12 pp.
25. **Hollis, D. G., R. E. Weaver, C. Baker, and C. Thornberry.** 1976. Halophilic *Vibrio* species isolated from blood culture. *J. Clin. Microbiol.* 3: 425–432; **Farmer, J. J.** 1980. Revival of the name *Vibrio vulnificus. Int. J. Syst. Bacteriol.* 30: 656.
26. **Blake, P. A., R. E. Weaver, and D. G. Hollis.** 1980. Diseases of humans (other than cholera) caused by vibrios. *Annu. Rev. Microbiol.* 34: 341–367; **Oliver, J. D.** 1981. The pathogenicity and ecology of *Vibrio vulnificus:* a particularly virulent microorganism. *Mar. Technol. Soc. J.* 15: 45–52; **Kaysner, C. A., C. Abeyta, Jr., M. M. Wekell, A. DePaola, Jr., R. F. Scott, and J. M. Leitch.** 1987. Incidence of *Vibrio cholerae* from estuaries of the United States west coast. *Appl. Environ. Microbiol.* 53: 1344–1348.
27. **Biosca, E. G., C. Amaro, C. Esteve, E. Alcaide, and E. Garay.** 1991. First record of *Vibrio vulnificus* biotype 2 from diseased European eel, *Anguilla anguilla* L. *J. Fish Dis.* 14: 103–109.
28. **Janssen, W. A.** 1970. Fish as vectors of human bacterial diseases. *Am. Fish. Soc.,* Spec. Publ. 5: 284–290.
29. **Lewis, D. H.** 1975. Retention of *Salmonella typhimurium* by certain species of fish and shrimp. *J. Am. Vet. Med. Assoc.* 167: 551–552.
30. **Cockey, R. R. and T. Chai.** 1988. An update on seafood-poisoning microorganisms. *Seafood Inf. Tips* 14(1): 1–6.
31. **Harada, M.** 1972. *Minamata Disease.* [In Japanese.] Iwanami Shoten, Tokyo, 274 pp.; **Huddle, N. and M. Reich.** 1975. *Island of Dreams: Environmental Crisis in Japan.* Autumn Press, New York. 225 pp.; **Smith, W. E. and A. M. Smith.** 1975. *Minamata.* Holt Rienhart Winston, New York. 220 pp.
32. **Kobayashi, J.** 1969. Investigations of the cause of itai-itai disease. I–III. [In Japanese.] *Kagaku* (Science) 39:286, 369, 424.
33. GESAMP. 1985. Review of potentially harmful substances — cadmium, lead and tin. *WHO Rep. Stud.* 22, Geneva; GESAMP. 1986. Review of potentially harmful substances — arsenic, mercury and selenium. *WHO Rep. Stud.* 28, Geneva; **Friberg, L.** 1988. The GESAMP evaluation of potentially harmful substances in fish and other seafood with special reference to carcinogenic substances. *Aquat. Toxicol.* 11: 379–393.
34. **Swain, W. R.** 1988. Human health consequences of consumption of fish contaminated with organochlorine compounds. *Aquat. Toxicol.* 11: 357–377; **Fein, G. G., S. W. Jacobson, P. M. Schwartz, and J. L. Jacobson.** 1981. Intrauterine exposure to polychlorinated biphenyls: effects on infants and mothers. University of Michigan, School of Public Health, Ann Arbor, MI. Unpubl., 215 pp.; **Fein, G. G., P. M. Schwartz, S. W. Jacobson, and J. L. Jacobson.** 1983a. Environmental toxins and behavioral development: a new role for psychological research. *Am. Psychol.* 38: 1188–1197; **Fein, G. G., J. L. Jacobson, S. W. Jacobson, and P. M. Schwartz.** 1983b. Intrauterine exposure of humans to PCBs: newborn effects. Final report. U.S. Environmental Protection Agency, Grosse Ile, MI. Unpubl., 54 pp.; **Fein, G. G., J. L. Jacobson,**

S. W. Jacobson, P. M. Schwartz, and J. K. Dowler. 1984. Prenatal exposure to polychlorinated biphenyls: effects on birth size and gestational age. *Pediatrics* 105: 315–320; **Humphrey, H. E. B.** 1976. Evaluation of changes of the level of polychlorinated biphenyls (PCBs) in human tissue. Final Report. U.S. FDA contract, Michigan Dept. of Public Health, Lansing, MI, 86 pp.; **Humphrey, H. E. B.** 1983a. Population studies of PCBs in Michigan residents, pp. 299–310. in D'itri, F.M. and M.A. Kamrin (Eds.), PCBs: *Human and Environmental Hazards*. Butterworth Publishers, Boston, MA; **Humphrey, H. E. B.** 1983b. Evaluation of humans exposed to waterborne chemicals in the Great Lakes. Final report. Environmental Protection Agency, Michigan Dept. of Public Health, Lansing, MI, 86 pp.; **Jacobson, J. L., G. G. Fein, S. W. Jacobson, P. M. Schwartz, and J. K. Dowler.** 1984a. The transfer of polychlorinated biphenyls (PCBs) and polybrominated biphenyls (PBBs) across the human placenta and into maternal milk. *Am. J. Public Health* 74: 378–379; **Jacobson, J. L., S. W. Jacobson, P. M. Schwartz, G. G. Fein, and J. K. Dowler.** 1984b. Prenatal exposure to an environmental toxin: a test of the multiple effects model. *Dev. Psychol.* 20: 523–532; **Jacobson, S. W., G. G. Fein, J. L. Jacobson, P. M. Schwartz, and J. K. Dowler.** 1985. The effect of intrauterine PCB exposure on visual recognition memory. *Child Dev.* 56: 853–860; **Jacobson, S. W., J. L. Jacobson, P. M. Schwartz, and G. G. Fein.** 1983. Intrauterine exposure of human newborns to PCBs: measures of exposure, pp. 311–343. in D'itri, F.M. and M.A. Kamrin (Eds.), *PCBs: Human and Environmental Hazards*. Butterworth Publishers, Boston, MA; **Schwartz, P. M., S. W. Jacobson, G. Fein, J. L. Jacobson, and H. A. Price.** 1983. Lake Michigan fish consumption as a source of polychlorinated biphenyls in human cord serum, maternal serum and milk. *Am. J. Public Health* 73: 293–296; **Wickizer, T. M., L. B. Brilliant, R. Copeland, and R. Tilden.** 1981. Polychlorinated biphenyl contamination of nursing mothers' milk in Michigan. *Am. J. Public Health* 71: 132–137.

35. **Butler, P. A.** 1969a. The significance of DDT residues in estuarine fauna, pp. 205–220. in Miller, M.W. and G.G. Berg (Eds.), *Chemical Fallout*. Charles C Thomas, Springfield, IL; **Butler, P. A.** 1969b. Monitoring pesticide pollution. *BioScience* 19: 889–891; **Butler, P. A.** 1973. Residues in fish, wildlife and estuarines: organochlorine residues in estuarine mollusks 1965–72 — National pesticide Monitoring Program. *Pest. Monit. J.* 6: 238–362.
36. **Cordle, F., R. Locke, and J. Springer.** 1982. Risk assessment in a federal regulatory agency: an assessment of risk associated with the human consumption of some species of fish contaminated with polychlorinated biphenyls (PCBs). *Environ. Health Perspect.* 45: 171–182.
37. GESAMP. 1983. Impact of oil and related chemicals on the marine environment. *Rep. Stud. GESAMP* 50, 180 pp.; GESAMP. 1985. The impact of carcinogenic substances on marine organisms and implications concerning public health. *GESAMP XV/2/4*. WHO Int. Rep., Geneva; GESAMP. 1986a. Impact of carcinogenic substances on marine organisms and implications concerning public health. *GESAMP XVI/2/3*. WHO Int. Rep., Geneva; GESAMP. 1986b. Occurrence of potential carcinogens and tumours in marine organisms. *GESAMP XVI/2/4*. WHO Int. Rep., Geneva; **Friberg, L.** 1988. The GESAMP evaluation of potentially harmful substances in fish and other seafood with special reference to carcinogenic substances. *Aquat. Toxicol.* 11: 379–393.

38. **Dunn, B. P. and H. F. Stich.** 1976. Release of the carcinogen benzo(*a*)pyrene from environmentally contaminated mussels. *Bull. Environ. Contam. Toxicol.* 15: 398–399.
39. GESAMP. 1985. The impact of carcinogenic substances on marine organisms and implications concerning public health. *GESAMP XV/2/4,* WHO Int. Rep., Geneva.
40. **Cabre, J. and J. Lasanta.**1968. Malignant epithelial tumors in deep-sea fishermen. *Actas Dermo-Sifiliogr.* 59: 361–364.
41. **Ayres, P. A. and M. Cullum.** 1978. Paralytic shellfish poisoning. An account of investigations into mussel toxicity in England 1968–1977. Ministr. Agr., Fish. Food, Dir. Fish. Res. Lowestoft, *Fish. Res. Tech.* Rep 40, 23 pp.; **Ayres, P. A., D. D. Seaton, and P. B. Tett.** 1982. Plankton blooms of economic importance to fisheries in UK waters 1968–1982. *Int. Counc. Explor. Sea,* Doc. C.M. 1982/L:38, 12 pp.
42. **Mee, L. D., M. Espinosa, and G. Diaz.** 1986. Paralytic shellfish poisoning with a *Gymnodinium catenatum* red tide on the Pacific coast of Mexico. *Mar. Environ. Res.* 19: 77–92.
43. **Ray, S. M. and W. B. Wilson.** 1957. The effects of unialgal and bacteria-free cultures of *Gymnodinium brevis* on fish and notes on related studies with bacteria. U.S. Fish and Wildlife Service, Special Science Report 211, 50 pp.; **Shimizu, Y.** 1983. Unexpected developments in red tide research. *Maritimes* 27(1): 4–6.
44. **Korringa, P. and R. T. Roskam.** 1961. An unusual case of mussel poisoning. *Int. Counc. Explor. Sea,* Doc. C.M.1961/41, 2 pp.
45. **Kat, M.** 1987. Diarrhetic mussel poisoning. Measures and consequences in The Netherlands. *Rapp. P.-V. Reun., Cons. Int. Explor. Mer* 187: 83–88.
46. **Pirquet, K. T.** 1988. Poisonous secrets. Shellfish testing in Canada. *Can. Aquacult.* 4(2): 41–43, 46–67, 53.
47. **Bird, C. J. and J. L. C. Wright.** 1989. The shellfish toxin domoic acid. *World Aquacult.* 20: 40–41.
48. **Waldichuk, M.** 1988. Mussel toxin leads to shellfish ban. *Mar. Pollut. Bull.* 19(3): 95–96.
49. **LoCicero, V. R. (Ed.).** 1975. Proceedings of the First International Conference on Toxic Dinoflagellate Blooms. 541 pp. Mass. Sci. Tech. Found., Wakefield, MA; **Taylor, D. L. and H. H. Seliger (Eds.).** 1979. *Developments in Marine Biology.* Vol. 1. *Toxic Dinoflagellate Blooms.* Elsevier/North-Holland, New York; **Spector, D. L. (Ed.).** 1984. *Dinoflagellates.* Academic Press, New York; **Anderson, D. M., A. W. White, and D. G. Baden (Eds.).** 1985. Toxic dinoflagellates. Proc. 3rd Int. Conf. *Toxic Dinoflagellates.* Elsevier/North Holland, New York, 561 pp.; **Okaichi, T., D. M. Anderson, and T. Nemoto (Eds.).** 1987. *Red Tides — Biology, Environmental Science and Toxicology.* Elsevier, New York; **Taylor, F. J. R. (Ed.).** 1987. *The Biology of Dinoflagellates.* Blackwell, Oxford; **Cosper, E. M., V. M. Bricelj, and E. J. Carpenter (Eds.).** 1989. *Novel Phytoplankton Blooms.* Springer-Verlag, New York; **Granelli, E., B. Sundstrom, L. Elder, and D. M. Anderson (Eds.).** 1990. *Toxic Marine Phytoplankton.* Elsevier, New York; **Smayda, T. J. and Y. Shimizu (Eds.).** 1992. *Toxic Phytoplankton Blooms in the Sea.* Elsevier, Amsterdam.

SECTION III:
Effects of Ocean Pollution on Marine Populations

The preceding section of this book focused grimly on diseases related to ocean pollution, as seen from resource as well as public health perspectives. Armed with that necessary background, it is time now to plunge into what should be the most critical pollution effects—those that affect population abundance, since it is here, in survival or death, that the most insidious long-term effects may be seen.

Decline in abundance of a population—any population—can result from increased mortality or decreased fecundity (or a combination of the two). Pollution does influence both processes. Severe pollution stress can kill individuals in localized areas of heaviest concentrations. Acute lethal effects can be recognized, often as local fish kills, and can even be quantified in some instances, especially in anadromous species, where effects of dams and pulp mill pollution have been examined for more than a century. Sublethal chronic effects of pollution can be reflected in decreases in fecundity, influenced by reduced growth or impaired metabolic pathways. Such sublethal effects, especially those that lead to reproductive failure, may be of great significance to long-term survival and abundance of stocks of marine fish, but unfortunately (as we will see in the following chapters), population reductions due solely to pollution have not been clearly demonstrated for many offshore marine fish.

The ways in which marine populations vary, and the environmental factors responsible for that variability, have been examined doggedly by quantitative biologists for more than a century. Some limited understanding has been developed, to the point where predictive models have been proposed and tested—with marginal success. We need much more survey data on how marine populations change in abundance, and more investigations of how environmental factors such as temperature, salinity, predation, competition, starvation (and pollution) influence abundance. But the need for more data is the inevitable scientific response to any environmental problem and we are not going to complain about its absence here. It seems more relevant to our discussion to examine some of the information that *does* exist about population fluctuations, and especially about the part that pollution may play in causing observed changes.

This is obviously not a simple topic, but it is one of great importance to any understanding of the role of marine pollution and the future of both wild and cultivated stocks, so we should attempt to confront it head on. I have elected to make the attempt by dissecting the topic of population effects of pollution into five subsections, with a chapter for each:

Mass mortalities in the sea—natural and pollution induced

Effects of pollution on fish abundance

Effects of pollution on shellfish abundance

Effects of pollution on aquaculture populations and

Global changes in the health of marine populations that may be associated with human activities (coral bleaching, sea urchin mortalities, fish ulcers, marine mammal mortalities, and toxic algal blooms).

This agenda is ambitious, and eventual integration of these subsections may prove to be elusive, but lurking here are the ingredients for better understanding of the quantitative effects of pollution on marine populations, if we can keep the narrative from descending into chaos.

6 Mass Mortalities in the Sea: Natural and Pollution-Induced

****

HERRING OF THE GULF OF SAINT LAWRENCE

Each spring, sometimes even before the ice has disappeared from the Gulf of Saint Lawrence, great silvery masses of sea herring, a species known technically as Clupea harengus, move westward into the Gulf to spawn off the coasts of the Magdalen Islands, Quebec, and northern New Brunswick. Following ancient rhythms, the huge aggregations crowd into shallow waters to reproduce, and then disperse to the east later in the season.

In some years though — at roughly quarter-century intervals — these spawning migrations become transformed into what seem like predestined appointments with death. The incoming herring schools encounter in those coastal waters large populations of a virulent fungal pathogen, Ichthyophonus hoferi. The spores of the fungus, present in nearshore waters, cause systemic infections in herring that are usually lethal within a few weeks. Fish die in truly overwhelming numbers; carcasses float on the surface, litter the bottom, or are washed up in windrows on extensive stretches of Gulf beaches. In peak mortality years, an estimated one-quarter of the entire adult herring population of that major body of water can be killed by the disease, and commercial catches are drastically reduced for several years following each outbreak.

Then, in a year or two after the peak of mortalities, just as mysteriously as it has appeared, the disease retreats. Infection rates decline rapidly and mortalities cease. Annual spawning migrations then go on for several decades unhindered by the threatening presence of fatal disease. Since the late 19th century, three outbreaks of this kind have occurred in herring stocks of the Gulf of Saint Lawrence, each one following a remarkably similar pattern.

This extraordinarily complex interweaving of major ecological events — spawning aggregations clouded by epizootic disease, reproduction in the immediate presence of mass deaths — illustrates the limited distance that we have traveled toward understanding mortalities in marine fish populations. We still have a very marginal comprehension of these disease-induced phenomena; we still stand bewildered on the shores — almost as did our ancestors from the dark ages, when confronted by the great human pestilences of those far-off times.

From "Field Notes of a Pollution Watcher"
(C. J. Sindermann, 1970)

****

INTRODUCTION

A marine scientist, Margaretha Brongersma-Sanders, looking for an explanation for petroleum deposits, published an exhaustive account 38 years ago of mass mortalities in the world's oceans — a massive review (70 pages of very small print) of every published event up to that time.[1] She considered physical, chemical, and biological causes of mortalities, but it is interesting that she had no category for epizootic disease or pollution-induced mortalities and made absolutely no mention of the role of disease or of pollution in the sea in all those 70 pages of fine print. Maybe this is understandable, since her background was in marine geology, but events have occurred in the past three and one-half decades that justify a reappraisal of mass mortalities in the sea — from a different but probably still biased viewpoint.

My thesis in this chapter is that there are new or badly underestimated mortality factors in the sea that may reduce marine populations — sometimes abruptly and dramatically enough to fit a definition of mass mortality. These include:

1. Epizootic disease
2. Large-scale marine events of physical-chemical causation
3. Lethal effects of stress from environmental pollution

We might begin by defining terms. "Mortality," of course, is a continuing phenomenon in any population — and is particularly dominant but not always obvious *early in the life history* of individuals comprising that population. Every biologist has been exposed in some way to population mortality curves, but acquiring data to prepare such curves for marine species is painfully slow and expensive; much extrapolation is done, and mortality is often (and incorrectly) considered by biometricians as a *constant* in assessing fish and shellfish populations.

Just as one example of the expression of this population curve for a marine fish, the winter flounder, Saul Saila, a distinguished professor at the University of Rhode Island, found that the average fecundity of winter flounder in New England was 600,000 eggs. Of these, *only 10% of spawned eggs hatched*. Mortality was highest in newly hatched larvae and declined with time, but only 18 individuals — 0.018% of the 10% that hatched — survived to age 1 from 100,000 newly hatched larvae.[2] He pointed out that "Determination of factors affecting mortality rate for any given year class becomes of great significance if departures from average or expected values (background values) occur, since recruitment to a fishery may be drastically affected." Mass mortality becomes one such disturbing factor.

A "mass mortality" can be described as "an unusual and sharply defined increase in mortality rate, of sufficient proportions to affect population size significantly and to at least temporarily dislocate the ecosystem of which the population is a part" (author's definition). Mass mortalities may be *local* — confined to a particular cove or estuary — or they may be *widespread,* affecting hundreds of miles of coastline.

Causes of mass mortalities can be *physical, chemical,* or *biological* — or combinations that produce stress beyond the limit of tolerance of individuals in the population.[3]

Physical: Vulcanism, temperature changes and extremes, storms, seaquakes, vertical upwelling currents
Chemical: Salinity changes and extremes, oxygen depletion, contaminant chemicals, hydrogen sulfide formation
Biological: Predation, disease, starvation, algal blooms, and toxins

This, you might be thinking, is all textbook stuff: what new insights have been gained? What additions to existing data have been made?

I'd like in this chapter to discuss some post–Brongersma-Sanders information about mass mortalities due to (1) *epizootic disease,* (2) *physical-chemical phenomena in the sea,* and (3) *pollutants*.

MASS MORTALITIES DUE TO EPIZOOTIC DISEASE

I did my doctoral thesis work a long time ago on an epizootic due to a systemic fungus infection in sea herring of the Gulf of Maine, a disease that closed the Maine sardine industry for most of an entire season. Then, shortly after I received my degree and a teaching job at Brandeis University, a much larger epizootic due to the same organism occurred in herring, alewives, and mackerel of the Gulf of St. Lawrence, truly a catastrophic event that I could not resist. That is when I began my long search to understand mass deaths in the sea, a search that continues to the moment.

I estimated from field observations made during that exciting period spent in Canada that over half of the herring and many of the mackerel of that entire Gulf were killed during the three-year outbreak period (1955–1957), and the immediate decline of the fishery to less than half its previous level supported the estimate.[4] Recovery was very slow. The epizootic was clearly caused by a virulent pathogen, and its course in the fish populations seemed to follow known epidemiological principles. The research (field and experimental) that I and Canadian colleagues did and the publications that resulted were very satisfying indeed, even though the fundamental causes of the outbreak remained obscure.[5]

We had scarcely finished our study in the Gulf of St. Lawrence when another and equally devastating epizootic began in oyster populations of the Middle Atlantic states — caused this time by an obscure protozoan pathogen known as "MSX." During the period from 1957 to 1963, most of the oyster stocks in Delaware Bay and lower Chesapeake Bay were killed. Mortality exceeded 80% in peak years.

Again, the epizootic was clearly caused by a virulent pathogen, but this time mediated by salinity and temperature. *This epizootic did not follow classical epidemiological principles,* and the present status of our understanding of the disease is less satisfying. We have never been able to transmit the disease experimentally, and infection rates remain variably high to this day in parts of Delaware Bay and Chesapeake Bay. Also, proximity to diseased oysters is not a critical factor in the spread of infection, suggesting a possible alternate host. Studies of this disease are still going on in New Jersey, Maryland, and Virginia.[6]

But we were diverted in 1963 by still another epizootic, this time in white perch and striped bass of the Chesapeake Bay caused by a species of the bacterial genus *Pasteurella*.[7] Again, significant numbers of fish died — enough to depress the white perch catches in the subsequent two years. Several groups studied the epizootic, and the consensus seemed to be that the pathogen was facultative and the fish population was under severe stresses of high density and high summer water temperatures. (A subsequent but smaller outbreak occurred in lower Chesapeake Bay in 1972.) Then, beginning in 1965 and extending to 1971, our attention was again diverted, when blue crabs began dying in coastal waters from Maryland to Georgia, causing a sharp reduction in catches. The pathogen this time proved to be an amoeba, *Paramoeba perniciosa*.[8] Mortalities were reported to reach 30% during sharp spring epizootic peaks.

Scientific attention lately — for the past two and one-half decades — has been on the coast of France, where so-called Portuguese oysters, *Crassostrea angulata* began dying in 1967, and European flat oysters, *Ostrea edulis* began dying in 1968. Much of the French native oyster industry was destroyed by 1975. The International Council for the Exploration of the Sea convened a working group of marine disease specialists from many countries to meet on the French coast in 1977 in an attempt (with French researchers) to identify the pathogens and suggest control measures. A viral disease was recognized in Portuguese oysters from the French coast, with the gills as primary focus of infection and with extensive tissue destruction — hence, the common name "gill disease".[9] Almost simultaneously, the pathogen thought responsible for mortalities of European flat oysters was identified as a haplosporidan protozoan, and a new genus *Marteilia* was created for the organism.[10] Primary effects of the pathogen were destruction of the digestive diverticular epithelium — hence, the common name "digestive gland disease". Mortalities began in a few small bays on the channel coast and spread to other growing areas on the Brittany Peninsula. Recently, an additional pathogen has been recognized — another protozoan, *Bonamia ostreae,* that has caused extensive mortalities in native flat oysters beginning in 1979 and continuing to the present time.[11]

The result of the combined mortalities was that culture of native European species on the northern French coast was virtually abandoned, and an exotic species, the Pacific oyster, *Crassostrea gigas,* imported as seed from Japan and British Columbia, has largely replaced the native oysters (well over 90% of production today).[12]

These are epizootics with which I have been personally involved, but this is of course only a very narrow sampling of the total world experience with recent disease-caused mortalities in the sea. Looking at the record of successes and failures in attempts to understand mass mortalities, it is not difficult to become a little discouraged. Often these "visitations" have appeared, exerted major impact on seafood production, then dwindled away, while individuals or small groups of scientists groped for causative factors.

LARGE-SCALE MORTALITIES CAUSED BY EXTREME PHYSICAL-CHEMICAL CONDITIONS

Next, I'd like to consider the category of large-scale marine catastrophes of physical-chemical etiology. While a number of such factors that have caused mass mortalities — such as vulcanism, storms, extreme temperature and salinity changes, contaminant chemicals, and oxygen depletion — were listed earlier in this chapter, some specific examples could be useful.

The effects of oxygen depletion on fish populations have, until recently, been badly underestimated. Those who are familiar with Baltic environmental problems or oxygen depletion in Norwegian fjords, will have already encountered the devastation possible from anoxia; nutrient enrichment of estuarine and coastal waters will undoubtedly result in greater familiarity with problems of low environmental oxygen levels, even in areas where they have not been known previously.[13] Our encounter in 1976 with severe and persistent oxygen depletion on the east coast of the U.S. off New Jersey left little doubt about the potential effects of this single factor on resource species.

Since this event was so extensive, covering 12,000 km^2; since it occurred on an open continental shelf; since human chemical inputs may have been contributory; and since oxygen depletion may be a much more common phenomenon than we think — it seems worthwhile to review, as a case study, the most important aspects of the 1976 event on the North American continental shelf.

Our involvement in this event began innocuously enough, with reports during the first week of July 1976 from sport divers, lobstermen, and trawler fishermen of dead and dying fish and invertebrates on fishing reefs and wrecks off the coast. Similar but much more localized events had occurred in 1968, 1971, and 1974. Within a few days, however, the reported mortality areas had extended southward some 70 km and well out on the continental shelf. We began a series of survey cruises to assess the damage — cruises that had to be extended farther and farther southward and seaward to find the boundary. Oxygen-deficient bottom water — sometimes with zero dissolved oxygen levels — was found in a zone with a coastal distance of some 165 km, in a corridor from 5 to 85 km off the coast. We knew then that we were in the presence of an environmental event of extraordinary proportions.

In the central coastal area, oxygen values near the bottom were zero, and hydrogen sulfide was formed. The anoxic condition persisted until October, when reduced surface temperatures and mixing gradually reoxygenated the bottom water.

Mortalities of fish, lobsters, and molluscan shellfish were observed. The sedentary forms — surf clams, ocean quahogs, sea scallops, and the benthic infauna — suffered the greatest mortalities. From our more or less continuous surveys, we estimated that at least 60% of the surf clam, *Spisula solidissima,* population off the central New Jersey coast — some 147,000 tons — had been destroyed by October, with significant but lesser mortalities of ocean quahogs and sea scallops. Lobster catches were reduced by almost 50% during the period.

Mortalities of reef-dwelling fish were observed; estuarine species did not perform their usual offshore migrations, and coastal migratory species such as bluefish made a major seaward diversion around the anoxic zone.

As you might expect, the man in the street and newspaper feature writers immediately jumped to the conclusion that ocean disposal of pollutants, particularly sewage sludge dumping, which at that time was going on at a grand scale 11 mi from the New Jersey coast, was responsible for this environmental catastrophe. Our studies suggested otherwise — that large-scale meteorological and oceanographic phenomena were involved in the production of the anoxic zone. Extensive data were assembled and a hypothesis was developed, centering on a combination of anomalous environmental events superimposed on a marginal coastal area. A dominant element in the sequence of events was a massive bloom of the dinoflagellate *Ceratium tripos* over much of the Middle Atlantic Bight, but particularly concentrated in the New York Bight. The bloom began in February and persisted at least until July.

The final synthesis was this: "If oxygen demand from a declining phytoplankton bloom is superimposed on an area (New Jersey coast) already characterized by reduced dissolved oxygen in an average summer; and if this organically rich oxygen-demanding water is sealed off in spring by the early onset of a thermocline; and if water mass movement is reduced to a minimum flow of bottom water — the ingredients of disaster to marine animals are present." This may prove to be one of the best-documented examples that we have of mass mortality in the sea and its short- and long-term impacts on resource species,[14] although anoxic events of equal proportions have been observed and reported from other coastal areas in the world, including Walvis Bay in southwestern Africa and the deeper basins of the Baltic Sea. Lesser events have occurred in Conçepcion Bay, Chile; Mobile Bay, Alabama; the Gulf of Trieste in the northern Adriatic Sea; and in localized areas of the New Jersey continental shelf affected by the 1976 event.

A principal conclusion that can be drawn from this event is that physical-chemical phenomena — in the absence of any indication of infectious disease — can clearly cause mass mortalities in the sea.

Another example of how extreme variation in the chemical environment can result directly in mortalities of shellfish resource species occurred in Chesapeake Bay in 1972. A major storm, tropical storm Agnes, dumped unprecedented amounts of fresh water on the bay and its tributaries. Salinities in that entire complex estuarine system, one of the most productive in the entire country, were depressed drastically for a period of several months during the summer and early autumn. Effects of the freshet conditions on fish and shellfish populations were examined by several marine research institutions located on the estuary.[15] Principal biological findings included the following:

- The entire biological community was disrupted to some extent, and effects were still discernible two years later.
- Molluscan shellfish were most severely affected. Oysters in the central and upper portion of the bay suffered almost 100% mortality, and over 70% of the softshell clams in those zones died from combined effects of low salinity and high temperature.
- Adult and juvenile fish tolerated the reduced salinities well, although many moved downstream or to deeper water. Effects on fish were considered minor and transient.
- Fish eggs and larvae were washed out of the tributaries that serve as nurseries. At the mouth of one major river, the Rappahannock, as many as 6.5 million fish larvae per hour were being carried seaward during the peak of the flood, as estimated from plankton collections.
- Effects of the flood on benthos and plankton were greatest in the normally higher salinity lower estuary, a zone inhabited by species relatively intolerant of low salinity.

The low salinity event of 1972 in Chesapeake Bay was certainly among the most thoroughly studied of its genre, and a book on the subject was published in 1976 by the Chesapeake Bay Research Consortium. The general conclusion was that despite some impacts on the oyster and clam fisheries, the fish component of the bay's ecosystem demonstrated great resiliency in its responses to severe salinity perturbation.

It is obvious that the three examples discussed so far — epizootic disease, anoxia, and freshets in estuaries — represent extremes of environmental change, with expected major impacts on populations. Most environmental changes are of much smaller dimensions, with correspondingly less dramatic effects on marine organisms. Minor salinity changes, temporary increases in predator abundance, slight reduction in food supply, winter temperatures slightly below long-term averages — all are examples of minor perturbations that can still act as subacute stressors and can be factors in the survival and well-being of individuals.

EFFECTS OF POLLUTION

Effects of natural phenomena, major and minor, considered in the previous sections constitute an important background that must be understood, and against which *the effects of pollution* can be discussed. A remarkable expansion of investment in studies of pollution effects has occurred during the past two decades. In this or any comparable research area, it is essential periodically to assess the current status of understanding and to identify strengths and weaknesses in the accumulated data.

Pollutants can exert lethal or sublethal influences at any point in the life cycles of fish or their food organisms. Figure 25 illustrates the life cycle of a pelagic coastal species, the Atlantic herring, *Clupea harengus*. It seems almost predictable that with so many potential points of impact throughout life cycles, populations of these and other fish in contaminated waters should decline and eventually disappear, yet this has not been observed to date, and overfishing appears to be a more critical factor for some species.

Although there is considerable evidence for very localized effects of pollution on populations and many experimental demonstrations of effects on individual fish, there is as yet little specific evidence of widespread damage to major fisheries resource populations resulting from coastal/estuarine pollution. There is some evidence that other factors, such as repeated year-class successes or failures, long-term shifts in geographic distribution, and overfishing, may cause changes in fisheries of great magnitude, which could mask any subtle pollution effects. At present, the effects of offshore pollution cannot be separated clearly from effects of the many other stressors to which fish populations are subject. Localized reductions in abundance may be demonstrable in coastal/estuarine waters, but species throughout their geographic range give indications of variations that seem due to factors other than pollution.

As an example, a review of the fishery resources of the U.S.[16] discussed on a species-by-species basis the catches of all-important marine fish for the period from 1950 to 1970. Included were a number of Atlantic species that might be expected to demonstrate some effects of coastal pollution. With few exceptions, according to the author, catches of these species were relatively

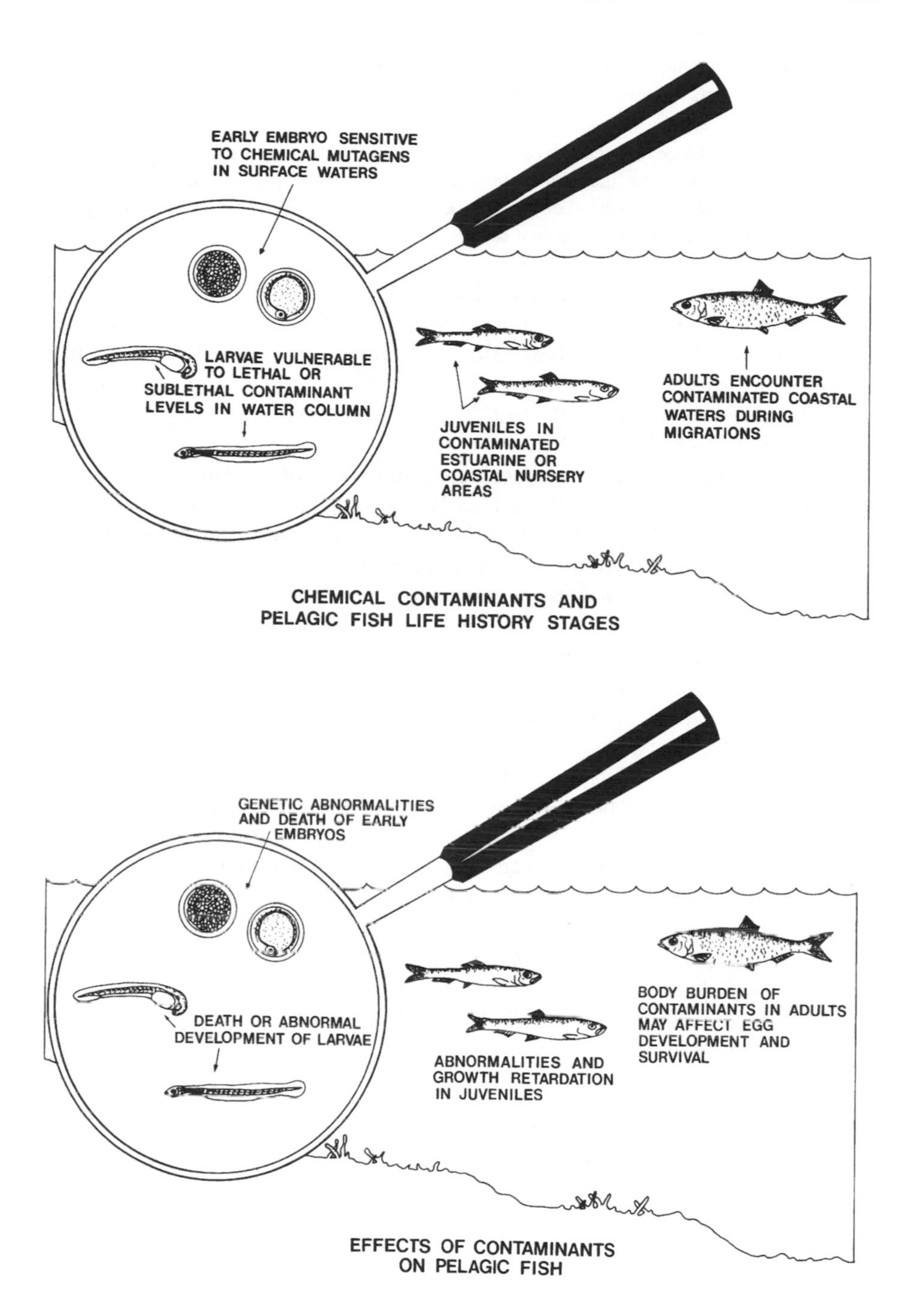

FIGURE 25. Vulnerable life cycle stages *(top)* and the effects of pollution *(bottom)* on Atlantic herring, *Clupea harengus*.

steady or showed some increases in the period from 1950 to 1970. One of the summarizing statements included in that report is of particular note:

> The evidence from catch records of a substantial number of exploited estuarine species indicates that pollution and damage to estuaries have not yet shown any measurable overall effects on the part of the marine resource which might be expected to show the first effects.

This statement discounts the possibility that certain anadromous or estuarine species may be resistant to at least some environmental changes and that species such as striped bass, *Morone saxatilis,* may actually benefit from certain man-made environmental alterations (such as increased turbidity). The statement also fails to take into account any significant changes in fishing effort or in species distribution.

As another example, a subsequent examination of fish landings from 1880 to 1980 in the heavily polluted New York Bight[17] failed to disclose consistent downward trends in species that, intuitively, would seem most vulnerable to damage from high pollutant levels in their habitats.

As a sweeping and therefore very dangerous generalization, there is no unequivocal evidence that fish species abundance has been affected by coastal/estuarine pollution, beyond limited but important localized effects. It should be kept in mind, however, that the demonstration of pollution effects on *individual* fish or on *local* populations should be sufficient cause for action by resource managers, just as any demonstration of pollution effects on individual humans is cause for action by regulatory and public health agencies. Corrective actions should not be deferred until major population effects can be demonstrated.

Another important perspective on the matter of pollution effects on marine populations was pointed out by a noted British marine scientist, Professor H. A. Cole:[18]

> ...pollution often results not in a gradual reduction in abundance of species but in a marked reduction in the variety of species without an accompanying reduction in biomass; indeed, among bottom organisms an increase in biomass may sometimes be observed.

According to this view, a search for obvious effects of pollution using the sole criterion of decrease in population size of any single resource species may obscure the larger ecological impacts.

CONCLUSIONS

To conclude then, I think there is room for updating our concepts of mass mortality in the sea. Certainly, epizootic disease has emerged as a major factor in recent decades. Experimental evidence garnered in the laboratory is accumulating about severe effects of pollutants, particularly on early life history stages of marine animals, and there are large-scale oceanographic phenomena that can have drastic effects on marine populations.

Fish population abundance is clearly influenced by a multiplicity of environmental variables. Man-induced chemical contamination of coastal/estuarine waters is one such variable, but possibly one that is dwarfed against the lengthened shadows of the many other factors, such as human predation, availability of food, and egg and larval predation, that affect survival. If there is validity to the conclusion that environmental changes influence survival and well-being of marine animals, then a conceptual base should be emerging from the many observations that have been made, and indeed a few concepts seem to be emerging. Some may seem almost like axioms, some may be erroneous, some may require modification, and certainly many have not even been formulated yet. Included for now are the following:

1. Mass mortalities of marine organisms due to extreme *natural* environmental changes occur, but normally they may constitute only a small part of the total damage to coastal/estuarine fish populations. Sublethal effects such as spawning failure, poor survival of larvae, reduced growth rates, and increased vulnerability to other environmental limiting factors, can have significant effects that may be less apparent and difficult to measure. Greater understanding of such sublethal effects is needed to assess fully the influence of any single stressor on living resources.
2. The so-called "resiliency" of marine populations is highly variable, since some mass mortalities are followed by very slow recovery of affected populations, whereas in other instances populations rebound rapidly. The amplitude of population fluctuations can be affected significantly by environmental factors that result in mass mortality. Also, drastic and sometimes long-term changes in community structure may be aftereffects of mass mortalities.
3. Success of reproduction and survival of year-classes depend on the persistence of a dynamic, even precarious, equilibrium — and man in some instances seems to be disturbing that equilibrium — by imposing additional environmental stressors.
4. Despite greater emphasis on marine studies, we are still often unable to distinguish with certainty between natural and man-made impacts on resource species.
5. At least some of the recent catastrophic events in coastal populations may have paleontological significance, because of the sheer numbers and the great areas involved, because environmental conditions could favor preservation, and because some events are repeated.
6. Intuitively, from results of experimental studies, it would seem that resource populations in contaminated waters should dwindle and disappear; yet, from our experience on the east coast of North America, this has not happened. Only in severely degraded local waters do we see localized disappearance of some species; and even in those areas other species that might be considered vulnerable are still present and are in some instances abundant. This is true particularly

of some of the coastal fish species, many of whom spend much of their life cycles in waters that are to some extent contaminated, especially the nursery areas for early life history stages. It may be that changes are occurring, but our assessments are inadequate to separate fluctuations due to natural causes from those due to pollution. It may be also that local populations are being augmented by migrations of stocks from less affected areas.

So, finally, Dr. Brongersma-Sanders, we have profited from your extensive early compilation of mass mortalities in the sea, but we would point out that in the years since you published your paper, evidence has accumulated that disease is in fact a major additional cause of mass mortality. Your paper, published almost four decades ago, serves to remind us, though, that disease is not the only answer; that physical-chemical factors beyond tolerance limits — temperature, oxygen, salinity — can kill too. The chemical factors now need to include those of human industrial origin as well. And even our augmented list of mortality factors may be incomplete; as an example, new evidence exists to suggest that increased radiation in Antarctic regions may be affecting productivity, an issue with potential global significance.

REFERENCES

1. **Brongersma-Sanders, M.** 1957. Mass mortality in the sea, pp. 941–1010. in Hedgpeth, J.W. (Ed.), *Treatise on Marine Ecology and Paleoecology,* Vol. 1, Chap. 29. Geological Society of America, New York.
2. **Saila, S. B.** 1962. The contribution of estuaries to the offshore winter flounder fishery in Rhode Island. *Proc. Gulf Caribb. Fish. Inst.* 1961: 95–105; **Berry, R. J., S. B. Saila, and D. B. Horton.** 1965. Growth studies of winter flounder, *Pseudopleuronectes americanus* (Walbaum), in Rhode Island. *Trans. Am. Fish. Soc.* 94: 259–264.
3. **Munro, A. L. S., A. H. McVicar, and R. Jones.** 1983. The epidemiology of infectious disease in commercially important wild marine fish. *Rapp. P.-V. Réun. Cons. Int. Explor. Mer* 182: 21–32.
4. **Sindermann, C. J.** 1956. Diseases of fishes of the western North Atlantic. IV. Fungus disease and resultant mortalities of herring in the Gulf of St. Lawrence in 1955. Maine Dept. Sea Shore Fish., *Res. Bull.* 25: 1–23; **Sindermann, C. J.** 1958. An epizootic in Gulf of Saint Lawrence fishes. *Trans. N. Am. Wildl. Conf.* 23: 349–360.
5. **Tibbo, S. N. and T. R. Graham.** 1963. Biological changes in herring stocks following an epizootic. *J. Fish. Res. Board Can.* 20: 435–449; **Sindermann, C. J.** 1963. Disease in marine populations. *Trans. N. Am. Wildl. Conf.* 28: 336–356.
6. **Ford, S. E. and H. H. Haskin.** 1982. History and epizootiology of *Haplosporidium nelsoni* (MSX), an oyster pathogen in Delaware Bay, 1957–1980. *J. Invertebr. Pathol.* 40: 118–141; **Andrews, J. D.** 1984. Epizootiology of haplosporidan diseases affecting oysters. *Comp. Pathobiol.* 7: 243–269; **Haskin, H. H. and S. E. Ford.** 1987. Breeding for disease resistance in molluscs, pp. 431–441. in Tiews, K. (Ed.), *Proceedings of the World Symposium on Selection, Hybridization, and Genetic Engineering in Aquaculture (1986).* Bordeaux, France.

7. **Snieszko, S. F., G. L. Bullock, E. Hollis, and J. G. Boone.** 1964. *Pasteurella* sp. from an epizootic of white perch *(Roccus americanus)* in Chesapeake Bay tidewater areas. *J. Bacteriol.* 88: 1814–1815.
8. **Newman, M. W. and G. E. Ward, Jr.** 1973. An epizootic of blue crabs, *Callinectes sapidus,* caused by *Paramoeba perniciosa. J. Invertebr. Pathol.* 22: 329–334; **Johnson, P. T.** 1977. Paramoebiasis in the blue crab, *Callinectes sapidus. J. Invertebr. Pathol.* 29: 308–320.
9. **Marteil, L.** 1968. La "maladie des branchies." *Int. Counc. Explor. Sea,* Doc. C.M.1968/K:5, 5 pp.; **Marteil, L.** 1969. Données générales sur la maladie des branchies. *Rev. Trav. Inst. Sci. Tech. Pêches Marit.* 33: 145–150; **Comps, M.** 1969. Observations relatives àl'affection branchiale des huîtres portugaises (*Crassostrea angulata* Lmk.). *Rev. Trav. Inst. Sci. Tech. Pêches Marit.* 33(2): 150–151.
10. **Grizel, H., M. Comps, F. Cousserans, J.-R. Bonami, and C. Vago.** 1974a. Étude d'un parasite de la glande digestive observé au cours de l'épizootie actuelle de l'huître plate. *C.R. Acad. Sci, Ser.* D 279: 783–784; **Grizel, H., M. Comps, J.-R. Bonami, F. Cousserans, J.-L. Duthoit, and M. A. LePennec.** 1974b. Recherche sur l'agent de la maladie de la glande digestive de *Ostrea edulis* Linné. *Sci. Pêche* 240: 7–30.
11. **Comps, M., G. Tige, and H. Grizel.** 1980. Étude ultrastructurale d'un protiste parasite de l'huître *Ostrea edulis* L. *C.R. Hebd. Séances Acad. Sci.,* Ser. D 290: 383–384; **Elston, R. A., C. A. Farley, and M. L. Kent.** 1986. Occurrence and significance of bonamiasis in European flat oysters *Ostrea edulis* in North America. *Dis. Aquat. Org.* 2: 49–54.
12. **Grizel, H. and M. Héral.** 1991. Introduction into France of the Japanese oyster *(Crassostrea gigas). J. Cons. Int. Explor. Mer* 47: 399–403.
13. **Boesch, D. F.** 1983. Implications of oxygen depletion on the continental shelf of the northern Gulf of Mexico. *Coastal Ocean Pollut. Assess. News* 2(3): 25–28.
14. **Swanson, R. L. and C. J. Sindermann (Eds.).** 1979. Oxygen depletion and associated benthic mortalities in New York Bight, 1976. U.S. Department of Commerce, NOAA Prof. Pap. 11, 345 pp.
15. Chesapeake Bay Research Consortium. 1976. The effects of tropical storm Agnes on the Chesapeake Bay estuarine system. CRC Publ. No. 34, Johns Hopkins University Press, Baltimore. MD.
16. **Wise, J. D. (Ed.).** 1974. The United States marine fishery resources. U.S. Department of Commerce, Contr. NOAA-NMFS MARMAP 1: 1–379.
17. **McHugh, J. L.** 1982. New Jersey fisheries — what is their future? U.S. Department of Commerce, Sandy Hook Lab., Highlands, NJ. NOAA-NMFS-NEFC Tech. Ser. Rep. No. 29: 7–53.
18. **Cole, H. A.** 1972. North Sea pollution, pp. 3–9. in Ruivo, M. (Ed.), *Marine Pollution and Sea Life. Fishing News Ltd.,* London; **Cole, H. A.** 1975. Marine pollution and the United Kingdom fisheries, pp. 277–303. in Harden-Jones, F.R. (Ed.), *Sea Fisheries Research.* John Wiley and Sons, London.

7 Effects of Pollution on Fish Abundance

****

MACKEREL MIGRATIONS IN WATERS OFF THE MIDDLE ATLANTIC STATES

The noted author Rachel Carson, in her first book Under the Sea Wind *published in 1941, wrote in almost poetic but still technically impeccable language about the annual spring migrations of Atlantic mackerel,* Scomber scombrus, *into the coastal waters of the Middle Atlantic States, and described in poignant terms the fates of billions of offspring produced during that mass movement — the many ways that death can and does come to far more than 99% of the helpless eggs and larvae in a hostile uncaring ocean environment.*

But Ms. Carson published her first book about the sea in a simpler time, well before toxic industrial chemicals were found to be so lethal to marine organisms, and more than two decades before she published her major work Silent Spring *(1962) that aroused public sensitivities to the menace of pesticides in the terrestrial environment. So in the spring of 1996, as in every other spring, the mackerel will again move in great masses over the continental shelves of the Middle Atlantic States. As they migrate, they will as usual produce uncountable numbers of offspring — except that now the beleaguered defenseless young will encounter an additional source of disability and death in the form of increasing levels of chemicals synthesized by man and discarded from his wasteful industrial processes. Developing embryos within the floating eggs and hatched larvae in the surface waters will encounter strange and often lethal man-made chemical creations — PCBs, DDT, polycyclic aromatic hydrocarbons, and many others — with which marine animals have had no previous evolutionary experience, and to which the younger life stages are particularly vulnerable.*

The precise percentage of all the mackerel offspring that will be disabled or killed this coming spring by the relatively new environmental stressor — pollutants at lethal levels — is not easily determined, but it is substantial, and the estimates are improving in quality. Some indications of the magnitude of mortalities have been obtained from percentages of dead eggs and abnormal embryos in samples that have been collected from coastal waters by the National Marine Fisheries Service intermittently since the mid-1970s. In one series of studies, a direct correlation was found between dead or abnormal young and environmental levels of contaminants throughout the New York Bight from Montauk, Long Island to Cape May, New Jersey. In some samples from severely contaminated inshore stations, as many as 40% of the eggs were found to be dead or to contain abnormal embryos. Fortunately, the spawning area for mackerel in the Middle Atlantic Bight extends well out into continental shelf waters, to areas where levels of contaminants are much lower than those in the immediate coastal zones, and where percentages of abnormalities are correspondingly smaller.

Scientists have learned much about mackerel during the past half-century. An extensive long-term data base, beginning in the 1930s, on size of mackerel stocks, spawning areas and intensities, and annual survival rates, can now be combined with more recent information on geographic distribution of pollutants in surface waters and prevalences of abnormalities and mortalities in eggs and larvae in those same areas. All these pieces of information provide an unusually strong opportunity for computer simulations of population impacts, and for realistic estimates of actual impacts of pollution.

Those estimates will not improve the odds for survival of any of the tiny vulnerable offspring from the 1996 spawning migration of mackerel in Middle Atlantic waters — but they will contribute to the gradual accretion of evidence that may someday convince a very destructive terrestrial species, Homo sapiens, to clean up its act, insofar as abuse of the oceans is concerned.

From "Field Notes of a Pollution Watcher"
(C. J. Sindermann, 1995)

****

INTRODUCTION

Most works of fiction have at least one passage that can be identified as "the climax." This is the point to which the rising action is directed and from which the falling action recedes. Nonfiction may or may not have such a critical narrative peak, but this book does (at least in the author's perception). The present discussion of quantitative effects of coastal pollution on resource populations should be it! The text thus far has been concerned with the biochemical/physiological effects of pollutants on survival and with the relationship of disease and coastal/estuarine pollution — both areas of inquiry that are important to any attempted assessment of pollution effects on the numbers of resource animals that survive or die.

We are as ready now as we ever will be in this document to confront the critical issue of numbers. It is not a topic that leads to crisp, satisfying conclusions, probably because it is so complex and our data sets are still so inadequate. To try to *isolate* and then to *quantify* specific pollutant effects on population abundance — as distinct from a maelstrom of other environmental influences on survival — is pushing the current state of the art in marine population dynamics too far. As will become obvious in the following pages, the approach that quantitative biologists usually resort to is to develop "simulation models", based on available data, to simulate and then predict potential effects of various levels of pollution (as additional causes of mortality) on the several life stages of fish and shellfish under varying conditions of exploitation by man.

The chapter has turned out to be a long and difficult one, but I hope it will prove to be satisfying for those who are really curious about the present state of understanding of pollution effects on *abundance* of marine species. I have tried a case history approach, in which several coastal/estuarine species at high risk from pollution have been examined in detail — made possible by the extent of available published information about those species.

So I offer this compilation with some trepidation; it may have too much of a textbook flavor, but there should be a few places in this book where we can try penetrating below the more easily accessible epidermis of a topic. This seems like one place to do it. The topic of quantitative effects of pollution can best be introduced by listing a series of hypotheses about the effects of pollution on fish and shellfish populations — hypotheses that have variable data support at present and may eventually be proved to be either correct or incorrect. They are:

- *Habitat destruction in estuaries and ocean disposal of contaminants have had localized adverse effects on resource organisms, but the overall effects at the species level have not been determined.* Lack of success in assessing effects of pollution on population abundance may be due in part to the fact that the *scale of impact* usually is much smaller than the geographical range of the species.

  Two decades ago, a British author[1] pointed out a critical guideline for evaluating conclusions reached about marine pollution: that we must be careful to differentiate large-scale changes due to pollution from purely local, almost parochial, situations, usually in a few square kilometers of coastal/estuarine waters.

  This admonition was augmented by other authors,[2] who stated that "Improvement in our capacity to predict the effects of pollution on fisheries requires an understanding of the environmental factors that control variability in fish populations and the effect of multiple stresses on these stocks over their entire geographical range. Increased predictive capability can be gained most cost-effectively through closer integration of the disciplines of population dynamics and toxicology."

- *Significant negative impacts of pollution on commercial fish stocks have not been demonstrated, even for areas such as the North Sea, where statistical information about fish stocks has been collected for many decades*. One British author,[3] for example, stated, "There is no evidence either way as to whether or not contamination of North Sea waters by metals, pesticide/residues, etc. has affected the well being of the fish stocks." This pronouncement may be correct, but it seems to be an oversimplification; a more rational viewpoint was expressed recently by German researchers,[4] who concluded that "...at present it is impossible to define the role of pollution on fish stocks of the North Sea as a whole. This is largely due to the fact that only drastic changes in marine ecosystems would be detectable and could be interpreted as manmade. Normally chronic and sublethal changes are taking place very slowly and it will be impossible to separate natural fluctuations from [those that are] anthropogenically caused. Long-term research over decades might be necessary to clearly distinguish between man-made and natural fluctuations especially in offshore waters."
- *Marine resource populations are subject to large natural fluctuations whose causes are incompletely understood*. Turning again to the North Sea, where resource and environmental data are probably better than almost anywhere else in the world, one noted German scientist's detailed report[5] on recent changes in fisheries and fish stocks concluded with this remarkably nebulous statement: "It seems that direct and indirect effects of changes in the fisheries as well as climatic changes and their consequences for the biotic environment caused the recent changes in the fish stocks of the North Sea. It is not possible to quantify the effects of man-made and natural factors separately because of the complexity of interactions between the various fish stocks and the stages of their early life history."
- *Because of the inability to distinguish population changes due to pollution, such changes might become catastrophic before they are noted*.

Effects of pollutants on reproduction and on survival until recruitment into the fished stocks may be particularly critical. Survival may be reduced by parentally transmitted or dietary contaminants, such as PCBs. Sperm viability and egg fertilizability may be reduced. Larval behavior may be affected, reducing competencies in food capture and predator avoidance. Pollution may result in increased susceptibility to disease in early life stages. The reproductive rate — usually considered by environmental toxicologists to be a more sensitive indicator than survival — may be severely affected. Toxic chemicals may reduce reproductive rates by causing diversion of energy to metabolic deactivation of harmful chemicals or to tissue repair; by directly impairing protein synthesis, hence growth; by affecting digestion and assimilation; or by interfering with gonad maturation and gamete production. These physiological responses of individuals under chemical stress can contribute to reproductive failure of the population as a whole, although other factors may be involved, as summarized by a Canadian author:[6]

"(1) the fish may be unable to reach its spawning grounds because of unfavourable ecological conditions, or its own weak physical state, and goes unspawned; (2) the eggs may never be released by the female owing to some unsuitable physiological condition; (3) the eggs and larvae may die because of their unhealthy state or poor conditions on the spawning grounds; (4) the eggs and larvae may be poisoned by a substance bioaccumulated in the gonads of the parent, e.g., DDT; or (5) the eggs and larvae may be poisoned by toxic substances in the environment."

- *Offshore fish stocks are not immune to damage from pollutants,* since year-class abundance is determined early in the life history, which for many marine species is spent in coastal/estuarine waters. The reality of this statement is demonstrated in the next major section on effects of pollution on menhaden stocks.
- *Many major fish stocks that are already overfished may be more vulnerable to additional stresses (such as pollution) than unexploited or lightly exploited populations.* A population that is heavily fished may have limited "compensatory reserve" (in the form of increased growth rates and greater egg production) and thus may be particularly sensitive to pollution (Figure 26). A critical question to be asked about each stressed population concerns this compensatory response, especially the amount of reserve that is left in an exploited population.

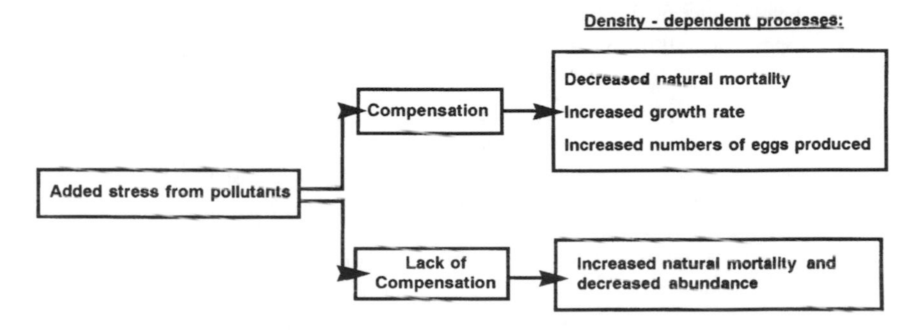

FIGURE 26. Population responses of pollution stress. Note that the gains from compensation are added (in an algebraic sense) to the losses from the added pollution stress. The net result could indeed be decreased *M*, but it could instead be constant *M* (if compensation precisely offsets the contaminant stress) or reduced *M* (if compensation is incomplete).

The point about differential effects of exploitation on stressed populations was made forcefully by a noted British researcher.[7] Defining the term *impact* as "the loss of eggs, larvae, or juveniles to the recruitment of fish stocks," he stated that:

> "Variability of recruitment in marine and anadromous fish stocks is high — one to two orders of magnitude — and an impact due to pollution would be difficult to detect. However, the effect of impact depends on the degree of exploitation, so if stocks are low and heavily exploited, no impact can be tolerated, until the stock has been returned to the desired objective, the maximum sustainable yield or any lesser quantity that might be optional."

This is consistent with a related statement about environmental effects by a contemporary British colleague[8]:

> "...in general, recruitment [annual addition to a fishable stock] is not directly dependent on the size of the spawning stock, but *appears to be dependent on other factors, frequently environmental in origin,* that affect the fish during the early larval and juvenile stages. Fluctuations in stock size are therefore very largely due to events that occur during the early life stages and are generally less dependent on events that influence the size of the adult stock."

Coastal/estuarine pollution can of course affect any life stage of fish (Figure 27), but it is during their first year of life — and more specifically during their first few months of life — that fish can be particularly sensitive to toxic contaminants. *Factors affecting early mortality are therefore of great significance in determining the causes of long-term population trends, even though death may occur at any point in the life cycle*. When translated into population terms, mortality may be chronic or catastrophic, as shown in Figure 28. Pollution impacts may occur at numerous points in this process.

Knowledge of the life history patterns of fish is important in determining the extent of pollution effects. Potentially critical considerations include:

- Location of spawning (fresh water, estuarine, coastal)
- Location of egg deposition (pelagic, demersal)
- Depth preference of hatched larvae in water column — surface film to bottom
- Location of nursery area for postlarvae and juveniles
- Feeding behavior and diets of all life stages
- Extent of migration into and out of polluted zones, and duration of occupation of those zones

These life history factors are of course modulated by the nature, extent, and intensity of pollution in every part of the habitat occupied by the species at any developmental stage, from embryo to adult.

CASE HISTORIES OF POLLUTION IMPACT STUDIES

Probably the most critical problem in assessing pollution effects on fish stocks is that of *separating natural and fishing mortality from pollution-induced mortality*. The nature of the problem was explored thoroughly by Jones,[8] whose illustration of the complexity of population responses is presented in Figure 29.

The extent of continuing frustration about inability to distinguish causation can be detected in extreme statements such as "...biological monitoring programs that cannot separate pollution-induced change from natural change

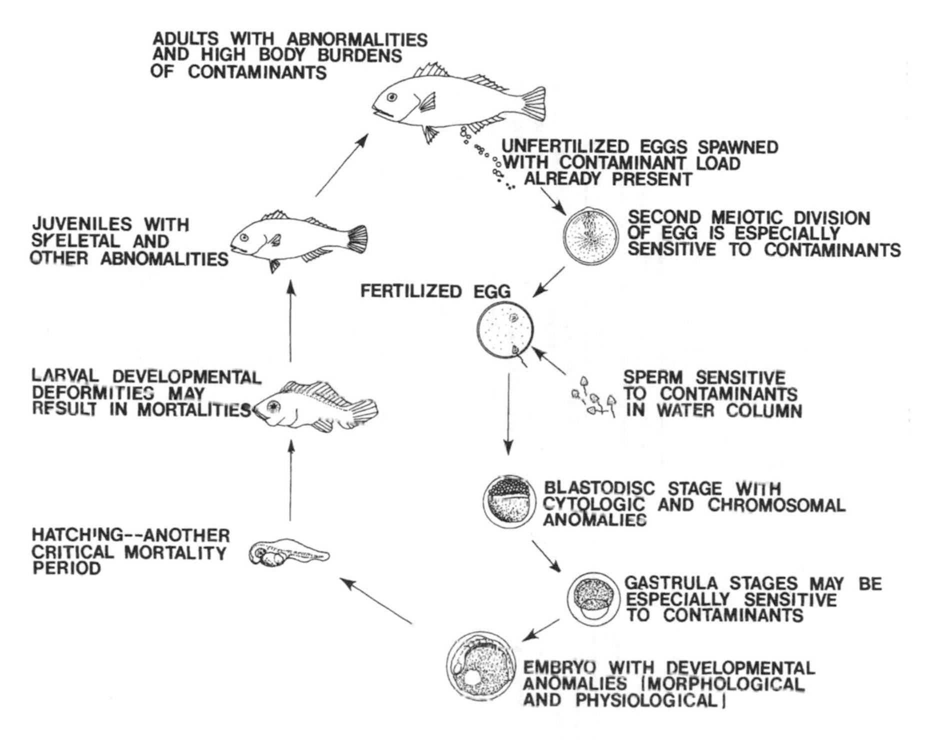

FIGURE 27. Points in the life cycle when fish are especially sensitive to pollutants.

should be terminated and those regulatory requirements which require such programs changed".[9] A reasonable rebuttal to such subjective conclusions might be (in my opinion) that we should intensify research leading to better delimitation of pollution induced from natural changes in marine populations, and in the interim should adopt a "precautionary principle" developed recently by Germany,[10] and accepted at the Second International Conference on the Protection of the North Sea in 1987. The principle *requires action to reduce pollution even in the absence of soundly established scientific proof for cause-and-effect relationships*. A proposed elaboration of the principle states that "Only those reduction measures will be applicable which are technically and economically feasible. But the decision whether xenobiotic substances are introduced into the marine environment or not should not be based on considerations of the assimilative capacity of the recipient water but on technically available reduction options." Additionally, the precautionary concept always has to be accompanied by intensive research on the effects of pollutants in the marine environment. The principle is a policy-making strategy, acknowledging

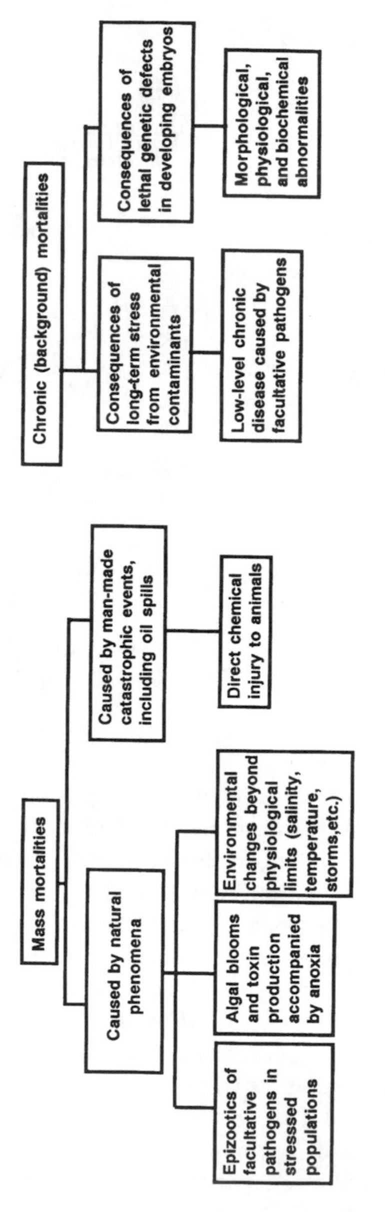

FIGURE 28. Causes of mortality in marine fish.

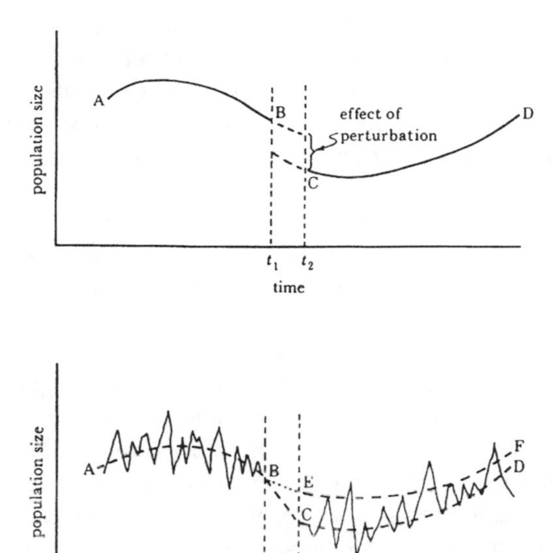

FIGURE 29. (*Top*) Simplified response in which a population subject to long-term cyclical variation at level AB is perturbed from $time_1$ to $time_2$, displacing the two parts of the cycle; and (*bottom*) response of a population subject to long-term cyclical variation and short-term random variation (AB) that is perturbed from $time_1$ to $time_2$. The result of perturbation causes a change in population size (BC) to be different than had the perturbation not occurred. (From Jones, R. (1982),[8] copyright The Royal Society. With permission.)

that scientific evidence is often inconclusive, but providing for action to protect the environment even without that elusive scientific certainty.

Cause-and-effect relationships — demonstrations of population impacts of pollutants — have been subjects of searches by many investigators, especially during the past two decades. The approach selected for this chapter is to consider a few species of economic importance that may be at risk from coastal/estuarine pollution and to use these as "case histories" illustrating the current status of technical information about pollution effects on abundance. The species chosen are Atlantic menhaden, *Brevoortia tyrannus,* striped bass, *Morone saxatilis,* and winter flounder, *Pleuronectes americanus*.

Atlantic Menhaden

Estimates of mortality rates in fish stocks from fishing and from natural causes can be made through the use of a variety of models developed for assessment and management purposes. Such models may be adapted to assessing the effects of pollution by integrating pollution mortality terms into them. The modified models can then provide estimates of varying pollution mortality on fish stocks at various levels of exploitation. Such a model has been used to demonstrate population responses of menhaden, *Brevoortia tyrannus*, a relatively short-lived species, to simulated pollution events.[11] The investigators who developed the model pointed out that all sources of mortality — natural, fishing, and pollution — at all life stages of a species throughout its range must be quantified for effective prediction of pollution effects. Using the extensive (>30 years) menhaden data base of the Beaufort (NC) Laboratory of the National Marine Fisheries Service, and imposing both one-time pollution-related catastrophic mortalities and chronic mortalities on one year class, the response of the entire population for the next 30 years was simulated.

For a simulated catastrophic event (an oil well blowout in coastal waters), and assuming a one-time 50% reduction in survival of 0-age-group menhaden, the researchers estimated that the total biomass would be reduced by about 12% over 30 years. Based on the fact that stocks are heavily exploited already, and thus may have little compensatory reserve, the simulation indicated a permanent reduction in stock size of 9% as a consequence of the event. For a comparable simulation of chronic effects, assumptions were made of a slow but continuous decline in estuarine water quality and a substantial increase in ocean dumping. The simulation predicted a 40% decrease in total population biomass in 30 years.

In a subsequent paper by the Beaufort scientists, estimates of acute and chronic pollution effects derived from simulation modeling were made for Atlantic menhaden as well as for seven other inshore species.[12] The study was designed not only to examine population effects, but also to compare the relative vulnerability of several marine fish stocks to pollution. Findings included these:

- "The modeled stocks responded to a simulated catastrophic event (a one-time 50% reduction in first year survival) by taking, on average, 10 years to equilibrate at 88% of preimpact abundance."
- Species that are impacted the most by coastal/estuarine pollution seem to stabilize most rapidly following acute stress.
- Stocks most susceptible to acute stress are even more susceptible to chronic stress for at least up to 20 years.
- Estimates of first-year survival are critical to any attempts at simulation modeling.
- Estimates of a species susceptibility to pollution stress (in terms of stabilization time after impact) can be made from life history data, including age-specific survival and fecundity rates.

As the biologists pointed out, simulations of this kind can provide estimates of the magnitude and time duration of pollution impacts that should be useful to resource managers.

In a related modeling effort exploring the possible effect of a single man-induced catastrophic event (such as an oil spill) on stock abundance in future years, the staff of the Beaufort Laboratory examined the effect of mass mortalities of young menhaden that occurred in 1984. The study was designed to test the ability to detect reductions in populations following acute pollution events, because of variability in young-of-the-year survival.[13] They concluded that "...a catastrophic loss to the Atlantic menhaden 1984 year class (e.g., >50% loss in abundance of the 1984 menhaden year class from the entire Atlantic coast) would have to occur to be detectable at reasonable levels of statistical power (e.g., >70% chance of detection), but more subtle reductions (e.g., <25% loss to the Atlantic menhaden 1984 year class from the entire Atlantic coast) would undoubtedly go undetected (e.g., chance of detection <12%). *Such difficulties in detecting reductions are typical of most fish stocks having comparable or larger inherent variability in recruitment or landings*."

These studies with menhaden in the 1980s were preceded by similar attempts to evaluate the ability to detect reduction in year-class strength of white perch, *Morone americana,* in the Hudson River.[14] General conclusions from the white perch analysis were that at least 20 years of data collection would be required to detect an actual 50% reduction in mean year-class strength and that annual fluctuations can mask major reductions in mean year-class strength.

Simulation models, developed for predicting impacts of pollution and other man-induced habitat changes, can be designed to do two things:

1. Simulate the annual effect on recruitment of young-of-the-year (YOY) fish into the adult population (YOY model)
2. Simulate the long-term effect of reduced recruitment on population size (life cycle model)

The YOY models predict either the number of YOY surviving to age 1 or the percent reduction in the YOY. These outputs from YOY models are then used in conjunction with life cycle models to predict long-term yields or adult population reductions.[15]

Development and application of simulation models such as these to predict the effects of pollution on fish stocks is of course only one item in the battery of approaches that can be and are being applied to sorting out the causes of mortality in resource species. Other sources of information are:

- Data from fisheries landings and from fisheries-independent surveys, useful in estimating stock size, recruitment, and mortality
- Data on physical/chemical changes in the oceans
- Data from chemical analyses of contaminant levels in the habitat and in fish tissues
- Data on physiological responses and biochemical transformations of contaminants by marine organisms
- Data from experimental exposures to single or multiple contaminants
- Data from experimental field exposures of resource species in polluted zones

The level of effort required for a program that includes all these elements listed would appear to be overwhelming, but problems of this complexity can be addressed with modern computer power and a major long-term commitment of research resources.

Striped Bass

The striped bass, *Morone saxatilis,* has had a long history of extensive fluctuations in catches (and presumably abundance); its center of abundance is Chesapeake Bay, with lesser centers in the Hudson and Delaware estuaries. During their life cycles, striped bass occupy a variety of aquatic habitats, from the lower fresh-water zones of spawning rivers to the open sea (Figure 30). For most of their lives, but especially during the first year, they are exposed to an array of chemical contaminants, as well as to the highly variable estuarine and riverine environments. It is generally accepted that survival during the first 60 days of life determines the size of a year-class (Figure 31). It is also well accepted that certain of the early life stages are particularly vulnerable to environmental toxicants.

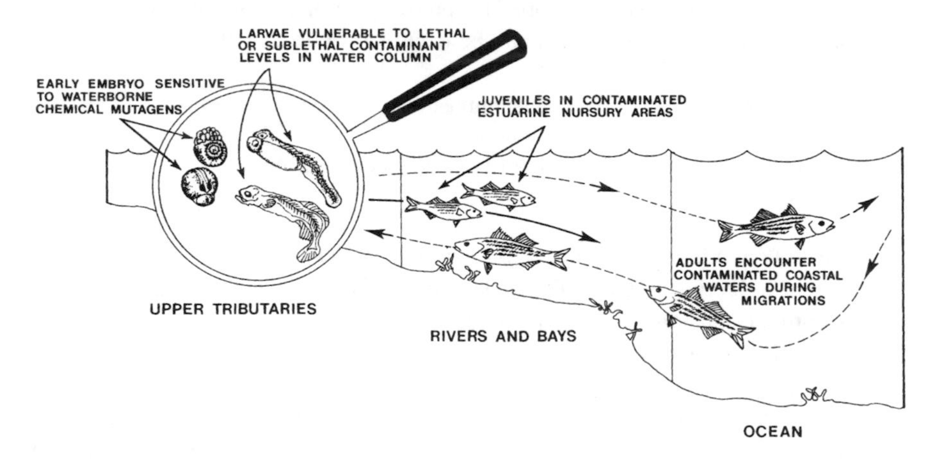

FIGURE 30. Life cycle of the striped bass, *Morone saxatilis,* with potential pollutant impact points.

Two aspects of human interference with the natural order of things (other than overfishing) have been responsible for extensive quantitative studies of striped bass on the Atlantic coast of the United States: (1) possible population impacts of power plant siting and operation on the Hudson River, and (2) the possible role of toxic contaminants in reducing abundance of Chesapeake Bay stocks. The Hudson River studies peaked in the mid-1970s; the Chesapeake Bay studies culminated in extensive documentation in the late 1970s and early

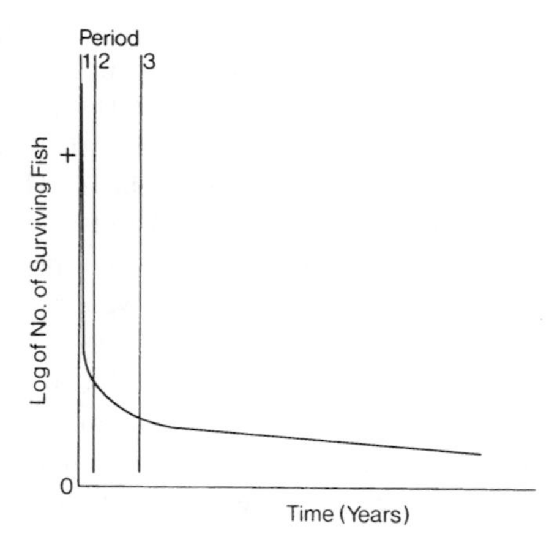

FIGURE 31. Hypothetical survival curve of one year-class of fish. The survival curve is divided into three periods: 1, eggs and larval fish; 2, prerecruit fish; 3, fished stocks. (Redrawn from Munro, A.L.S., A.H. McVicar, and R. Jones, *Rapp. P.-V. Reun. Cons. Int. Explor. Mer,* 182: 21, 1983, copyright ICES. With permission.)

1980s. Investigators were particularly interested in quantitative information linking human activities with the effects on striped bass survival.

The most recent decline in striped bass stocks in the Chesapeake Bay, beginning in 1973 and continuing to 1985, was attributed to various causes, especially overfishing and reduced survival of larvae because of chemical pollution. A major research effort in the late 1970s and early 1980s confirmed that fishing effort had increased since the mid-1960s and might have affected recruitment.[16] Evidence was also found of elevated levels of PCBs and other contaminants in tissues of young-of-the-year fish and in their habitats, and experimental exposure of yolk-sac larvae in the laboratory to environmental levels (but artificial mixtures) of selected contaminants resulted in increased mortality.[18] Yolk-sac larvae exposed in *in situ* chambers to Nanticoke River water (a tributary of Chesapeake Bay) also died differentially, and high aluminum and low pH were thought to be implicated. However, a later experimental study of yolk-sac larvae, also using *in situ* chambers placed in their natural habitat in the upper Chesapeake Bay (Chesapeake and Delaware Canal), did not indicate acute harmful effects, although sublethal gill abnormalities were seen in yearlings.[19]

In addition to the possible effects of toxic chemicals on larvae, effects on reproduction were also investigated as a possible cause of the population decline. Concentrations of PCBs up to 26 ppm were reported in adult striped bass from the Hudson River, and elevated concentrations of PCBs, DDT, and dieldrin were found in striped bass eggs.[20] The organochlorine residues were associated with the failure of cleavage in fertilized eggs, but correlations were not considered significant. In an earlier study of striped bass in California, eggs were found to contain 5 to 10 ppm DDT, but reproductive depression was not demonstrated.[21]

In other studies, the survival of larvae from eggs spawned by contaminated female striped bass was found to be inversely related to concentrations of chlorinated hydrocarbons (hexachlorobenzene, DDT, PCBs, and chlordane) in the eggs. It was also found that parental sources of these contaminants had greater effects on survival than did dietary sources. However, in an earlier investigation, eggs with PCB content from 1.1 to 8.1 ug/g (ppm) wet weight did not differ from controls in survival and growth after yolk absorption.[22]

In still other research, tissue contaminants were found to produce abnormalities that might affect stock abundance. Young-of-the-year striped bass from the Hudson River had high levels of PCBs in their tissues and had vertebrae that were fragile and that ruptured under minimal force.[23] The authors referred to other laboratory studies indicating that contaminants such as PCBs, cadmium, and lead could weaken vertebral structure and contribute to mortality.

Simulation models were used to examine the relative influence of two factors — increased fishing mortality and contaminant toxicity — in producing the observed decline in striped bass abundance.[24] Principal findings were:

- "...at low levels of density-dependent mortality, an increase in fishing mortality, or an equivalent decrease in early life-stage survival caused by the toxic effect of a contaminant would cause similar declines in the stock."
- "At high levels of density-dependent mortality, the effects on the yield are similar if the contaminant-induced mortality precedes the density-dependent mortality. However, if the contaminant-induced mortality occurs after the period of density-dependent mortality, the decline in yield will be more severe than that caused by an equivalent increase in fishing mortality."

However, despite the impressive amount of data and analyses available from this and other studies, the investigator concluded that "...the actual level of any excess mortality that is imposed on the striped bass population from toxic substances is unknown, and it will probably remain so for some time." But he then went on in a more optimistic vein, stating that reduced fishing mortality could result in a 20- to 30-fold increase in population fecundity which could "offset even rather severe losses due to contaminant toxicity, and could halt or reverse the decline in stock."

It is very interesting, and no doubt significant, that conclusions based on this modeling effort were supported by the behavior of the fishery and the population in the late 1980s. Beginning in 1985, the states of Maryland and Virginia imposed a moratorium on striped bass fishing that remained in effect until 1990. A dramatic increase in spawning stocks occurred (an estimated fivefold increase in spawning females in 1989 as compared to 1984), as well as a high average young-of-the-year index. Based on available evidence for the resurgence of striped bass stocks, the Atlantic States Marine Fisheries Commission (ASMFC) adopted a conservative management plan for limited harvest and a quota system beginning in 1990.

What can be concluded from the results of this research on Chesapeake Bay striped bass, followed by implementation of management measures and the resurgence of the population in the late 1980s? It begins now to appear that high fishing mortality was the principal culprit in the decline in stocks since 1973. (It might be noted parenthetically that to some quantitative biologists it has appeared that way all along, especially because fishing mortality rates for immature striped bass were very high.) It also appears that high larval mortality in some spawning tributaries may have been contributory, if stressful environmental conditions (pH, hardness) occurred coincident with toxic levels of specific contaminants (as, for example, aluminum did in the studies conducted in 1985 in the Nanticoke River). These conclusions support the concept that a heavily exploited population may have limited compensatory reserve and may be particularly sensitive to pollution. In the present case of striped bass, a drastic reduction in fishing mortality during the period 1985 to 1990 because of the moratorium allowed population expansion, even if other factors, such as low larval survival due to pollution, had contributed to the previous decline.

WINTER FLOUNDER

Over the past two decades, the winter flounder, *Pleuronectes americanus,* has been the subject of many studies that have emphasized pollution effects on early life history stages (Figure 32). The average female produces about 600,000 eggs, of which an estimated 10 to 16% hatch. Only an estimated 18 individuals per 100,000 hatched larvae survive to age 1.[25] The species is estuarine-dependent, both for nursery areas and for overwintering sites for adults. Many estuaries are polluted, and much of the life history of the winter flounder is spent in close association with contaminated bottom sediments in those estuaries. Adults lie partially buried in bottom sediments; spawning occurs near the bottom; eggs sink to the bottom, where they aggregate in clusters; and larvae, after hatching, alternately swim upward and then sink to the bottom.[26] Fin erosion (a good pollution indicator) is a condition commonly seen in adults from polluted habitats, and tissue levels of pollutants in juveniles and adults can be significantly elevated.

The accumulation of chlorinated hydrocarbons in the tissues of adults and their transfer by females to eggs has been found in several studies to contribute substantially to larval mortality. Evidence was found in studies conducted in the 1960s that high mortalities of larval winter flounder in a Massachusetts estuary (a tributary of Buzzards Bay) could be related to pesticide pollution.[27] Adult females concentrated DDT, DDE, and heptachlor epoxide in their ovaries as spawning approached, and mortality of post-yolk-sac larvae was estimated to approximate 100%. The authors pointed out the similarity of this pattern of reduced hatchability and larval mortality to that reported for several salmonid species, and considered it to be the result of DDT contamination.[28] The

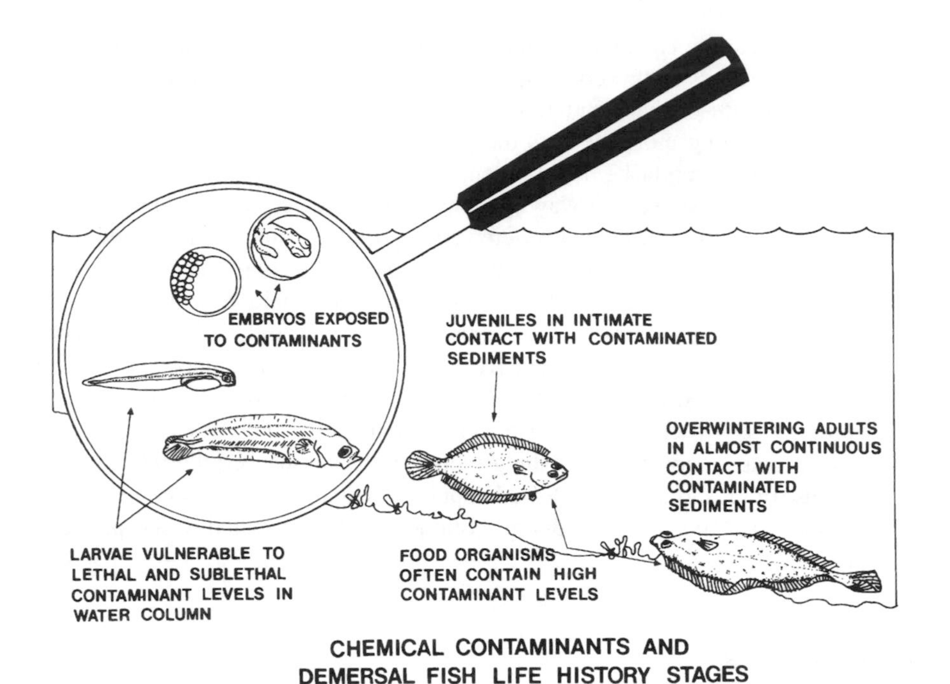

FIGURE 32. Life history stages of demersal fish that are vulnerable to pollution.

explanation offered was that DDT was translocated to the maturing eggs, where, bound to yolk fats, it remained inactive biologically until such fats were metabolized by the developing fry; DDT was then released with lethal results.

Corollary to this study of mortality of winter flounder larvae, juveniles of age 2 and younger, year-round residents of the same polluted Massachusetts estuary, contained higher tissue levels of pesticide residues than did the migratory adults, but mortalities were not observed.

In a related experimental study of DDT effects on developing eggs, adult female winter flounders were exposed to sublethal concentrations of DDT and dieldrin. Dieldrin exposure did not affect survival to hatching,but DDT exposure of females resulted in abnormal gastrulation and mortality of eggs after fertilization, and severe vertebral deformities in 39% of the larvae at hatching. Experimentally induced gonad levels of the insecticides in spawning females duplicated levels found in wild fish.[29]

More recent studies of the effects of polluted habitats on the reproductive success of winter flounder have produced conflicting results. Examination of samples collected along a composite pollution gradient in Long Island Sound found no significant differences in the percentage of viable hatch from eggs taken from females at each site. In cytogenetic studies, however, increased percentages of chromosomal anomalies and reduced mitotic rates characterized embryos from the more polluted sites (western Long Island Sound).

Analysis of PCBs in eggs did not indicate a correlation between contaminant levels and the cytogenetic data.[30]

Another pollution gradient, this one a composite of stations in Narragansett Bay, Rhode Island, and Buzzards Bay, Massachusetts, was exploited in a study of the effects of inherited contaminants on eggs and larvae of resident winter flounder.[31] Progeny from flounders captured in Buzzards Bay, an area noted for long-term PCB contamination, contained significantly higher levels of PCBs (averaging 39.6 μg/g dry weight), and hatched larvae were smaller in length and weight than progeny from reference site adults. No information was given on the percentage of viable hatch, but as larvae grew to metamorphosis the length/weight differences disappeared; the compensatory growth of larvae from contaminated parents but grown in clean water was attributed by the authors to biotransformation and detoxification of contaminants via mixed-function oxidase systems of the embryos and larvae.

Petroleum contamination can affect the survival of winter flounder larvae. Experimental exposures of mature female winter flounders and their developing eggs and larvae to low concentrations of No. 2 fuel oil produced results that should be useful in population analyses. Exposure to 100 ppb throughout the gonad maturation of parents, and during fertilization and embryogenesis, resulted in a three- to nine-day delay in hatching, a 19% reduction in viable hatch, and a 4% prevalence of spinal defects in hatched larvae. Larvae produced from gametes contaminated during parental gonadal maturation but then reared in clean water had a mortality coefficient of 0.130, considered by the authors to be much higher than the calculated mortality coefficient for untreated, laboratory-reared winter flounder larvae of 0.036 to 0.059.[32] Growth of larvae hatched from gametes from oil-exposed spawners was also slower. [It might be noted here that the topic of acute and chronic oil pollution in the sea was examined in detail from a fishery perspective by McIntyre (1982).[33] He concluded a detailed review by stating that "no long-term adverse effects on fish stocks can be attributed to oil, but local impacts can be extremely damaging in the short term...."]

The Northeast Fisheries Science Center of the National Marine Fisheries Service (Milford (CT) Laboratory) is carrying on an extensive interdisciplinary study of the effects of pollution on winter flounder populations of Long Island Sound. Among the recent findings were these:

- Fish from a severely polluted site had a smaller proportion of females actually spawning (whether naturally or artificially induced), a smaller proportion of live eggs at the time of extrusion, and a much smaller proportion of successful egg cultures than fish from a relatively clean reference site.
- The same polluted site had the highest level of developmental abnormality and mortality in the early life stages.
- In decreasing order of importance, pesticides, aromatic hydrocarbons, and polychlorinated biphenyls are the chemical body burdens that most limit the reproductive success of female winter flounder.

- In laboratory experiments, young-of-the-year winter flounder held under constant low levels (2.2. mg/l) of dissolved oxygen for 11 to 12 weeks grew only about half as much as flounder held under high levels (6.7 mg/l).
- Yearling winter flounder died during a 20-hour exposure at 20°C to a dissolved oxygen (DO) range of 1.1 to 1.5 ppm. They withstood, however, an eight-hour exposure to a DO range of 1.2 to 1.4 ppm at 20°C.

A three-year (1986–1988) study by Milford Laboratory staff members of winter flounder from selected stations in Long Island Sound and Boston Harbor disclosed a low percentage of viable hatch, small larvae, and delayed embryogenesis in offspring from fish taken at the most polluted sites (New Haven Harbor and Boston Harbor). Such findings indicate low larval survival and reproductive impairment at the most heavily degraded sites. A related study found the greatest prevalences of chromosomal abnormalities and mitotic disruptions in developing embryos from spawning adults taken at the most polluted sites.[34]

Results from earlier studies with estuarine-dependent species other than the winter flounder have provided additional evidence that high tissue concentrations of chlorinated hydrocarbons in spawning adults can result in mortalities of developing eggs and larvae. Reproductive failure of a sea trout, *Cynoscion nebulosus,* population in Texas was attributed to this phenomenon.[35] The sea trout population inhabited an estuary that was contaminated heavily with DDT, where DDT concentrations in ovaries reached a peak of 8 ppm prior to spawning compared to less than 0.5 ppm in sea trout from other less contaminated estuaries. Spawning seemed normal, but eggs failed to develop.

As a source of additional evidence of damage from chlorinated hydrocarbons, the reproductive success of starry flounders, *Platichthys stellatus,* from polluted San Francisco Bay was compared with that of a reference population from an unpolluted site. The total PCB content of eggs correlated inversely with embryological success and hatching success, supporting the stated hypothesis that chronic contamination of reproductive tissues by relatively low PCB concentrations (<200 μg/kg) has a pervasive deleterious effect on the reproductive success of starry flounders in San Francisco Bay.

Good evidence also came from European studies, in which Baltic flounders, *Platichthys flesus,* with elevated levels of PCBs in their ovarian tissues were found to have a significant reduction in viable hatch of larvae.[36] A threshold level of 120 ng/g (0.12 ppm) PCB (wet weight) in eggs and ovarian tissue was considered to be a contamination point above which reduced survival of developing eggs and larvae of that species could be expected. Levels of other chlorinated hydrocarbons or heavy metals could not be correlated with reductions in viable hatch. In a subsequent study of North Sea whiting, *Merlangius merlangus,* the same research team concluded that 0.2 ppm PCB in ovarian tissue constituted a threshold above which impaired reproductive success could be expected.

Effects of PCBs and DDE on reproductive success of Baltic herring, *Clupea harengus,* were also investigated.[37] Findings included these:

- Viable hatch was significantly reduced by ovarian PCB concentrations of more than 120 ng/g^{-1} and by DDE concentrations of more than 18 ng/g^{-1} (wet weight).
- A positive correlation existed between ovarian residues of PCBs and DDE.
- A linear relationship existed between ovarian residue levels of PCBs and DDE and viable hatch.
- The effects of PCBs and DDE on reproductive success were probably additive.

Levels of contaminants that reduced reproductive success in this study were low; the authors cautioned that other contaminants, not analyzed, may also be involved.

Thus far in this case history on winter flounder (augmented by data from other species), the effects of pollution on early life history stages have been emphasized, but effects on adults should not be ignored. Detailed studies of pollution effects on winter flounder began 25 years ago, with one of the first experimental studies of tissue lesions resulting from exposure to copper.[38] Principal effects were hemolytic anemia, fatty degeneration of the liver, and renal necrosis — all potentially lethal. Sublethal physiological effects of other heavy metals, such as cadmium and mercury, were reported by other investigators, and the distribution, metabolism, and excretion of DDT and Mirex were examined, with the interesting observation that the winter flounder stores its pesticide burden primarily in body muscle.[39]

A recent study of contaminants in winter flounder from a number of polluted sites on the northeast coast of the United States disclosed that levels of polycyclic aromatic and chlorinated hydrocarbons in stomach contents were higher than those in bottom sediments, indicating that the compounds were being accumulated by prey organisms (Figure 33).[40] Relating these findings to biological effects, the investigators described several kinds of necrotic and degenerative liver and kidney lesions that were relatively more prevalent in fish from severely polluted sites. Inferences were not made about possible pollution-induced mortality, but the correlation of certain sublethal pathological conditions with contaminated habitats and elevated contaminant levels in tissues is suggestive.

Related experimental studies, exposing adult winter flounders to oil-contaminated sediments for four to five months, resulted in mortalities in summer months, possibly because the oil was acting as a nonspecific stressor at a time when temperatures approached an incipient lethal level.[41]

The association of progressively severe liver pathology and several types of liver tumors with badly degraded coastal/estuarine waters is becoming apparent from recent reports. Among them, winter flounder from several degraded areas on the east coast of the U.S. (New Haven Harbor, upper Narragansett Bay, Boston Harbor) had prevalences of 3.4 to 7.5% tumors.[42] What seems to be emerging from examinations of tumors in winter flounders

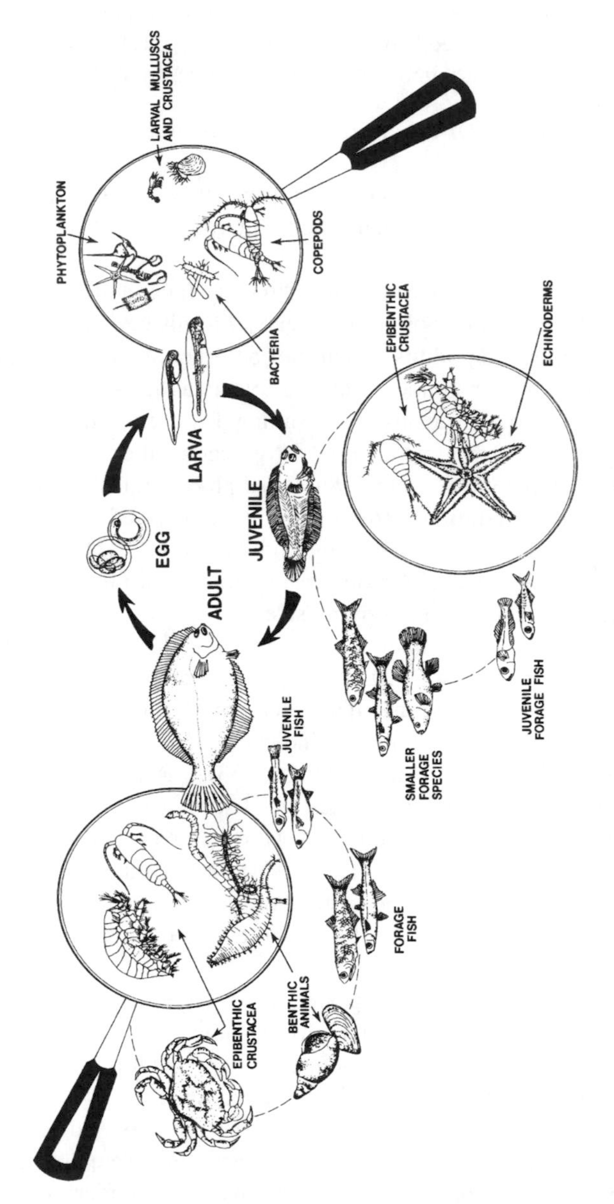

FIGURE 33. Effects of contaminants on food chains of winter flounder life history stages.

and other species in different polluted estuaries, is a sequence of histopathological changes in livers, beginning with fatty deposits and preneoplastic changes in liver parenchyma cells. The progression of pathological changes also seems roughly correlated with the extent of estuarine degradation and the length of residence of fish in the estuary. In studies of West Coast flatfish species, positive correlations were obtained between neoplasm prevalence in bottom-dwelling fish and levels of "certain individual groups of sediment-associated chemicals" (aromatic hydrocarbons, chlorinated hydrocarbons, and heavy metals).[43] Other studies of East Coast winter flounder have identified an oncogene derived from tumorous liver and possibly indicating a specific interaction of flounder DNA with polycyclic aromatic hydrocarbons (PAHs).[44]

A persistent question, still not resolved, is whether the liver tumors kill the fish or if they regress when the fish moves from the heavily polluted habitat. If the tumors do not regress, the survival of individual fish, as described by one investigator,[45] "depends on many factors, including the extent of liver damage, the degree of toxicity resulting from decreased hepatic function, the impairment of other essential organs, whether metastasis occurs, and the degree of behavioral modification (the ability to capture prey and avoid predators may be compromised). Population effects are possible if large numbers of fish with hepatic carcinoma die..."

Limited and somewhat conflicting recent information is available about a possible reduction in immune responses and hence a reduced survival potential of winter flounder as a consequence of exposure to pollutants. In one study, a four-month exposure to oiled sediments resulted in a statistically significant reduction in pigmented liver macrophage aggregates, believed to be important components of the cellular immune system of fish, and possibly primitive analogues of the mammalian lymph nodes.[46] However, in an earlier investigation,[47] the numbers of aggregates in fish from polluted areas were not greater than those from reference sites, although their size was. In still other studies with different fish species, aggregates were found to be more numerous and larger in samples from polluted areas than in those from unpolluted sites.[48] It seems plausible, as some investigators have pointed out, that at low chronic levels of pollution the cellular defense system may function effectively, whereas at higher, more toxic levels of pollution, phagocytosis may be impaired, leading to a decrease in melanomacrophage aggregates.

One more study of effects of pollution on adult winter flounders — this one demonstrating synergism between chronic oil pollution and protozoan (trypanosome) parasitization — was reported.[49] Infected fish exposed for six weeks to oil-contaminated sediments had higher mortality rates than uninfected individuals; intensity of infection was higher in oil-exposed fish than in untreated controls; and retardation of gonad development was more pronounced in the oil-treated, parasitized fish than in the other experimental fish.

So here then, with winter flounder, we have a species that has been examined extensively, from pathological, immunological, physiological, and biochemical perspectives, in an effort to understand effects of pollution. Probably the most significant findings, from a population point of view, are the demonstrated negative impacts on larval survival of exposure to chlorinated hydrocarbons, acting either on prespawning females or on eggs and larvae. Next in order of significance might be the severe morphological changes — liver tumors and fin erosion in particular — that have been seen in samples from badly degraded habitats such as the New York Bight apex, western Long Island Sound, and Boston Harbor. Not to be ignored, however, is the dissenting observation by one team of investigators[50] that chronically polluted environments seem to have little influence on winter flounder populations.

The one missing ingredient in this scrutiny of effects of pollution on the species is a serious attempt to *quantify* the observations made and *to provide numerical estimates of the extent of population reduction that may result from exposure to pollutants*. Some narrowly focused efforts have been made, such as Smith and Cole's[29] study of the effects of DDT on larval mortality, but we are left with little published information that can be extrapolated to the entire population of winter flounder on the northeast coast of the United States.

At present, an initial effort is being made to develop and test a model to assess effects of pollution on winter flounder population abundance in Long Island Sound (F. Thurberg, personal communication).* The sheer mass of relevant information available — biological and environmental — should enhance the likelihood of success for such a project.

In advance of the availability of any model, it might be instructive to discuss the kinds of information that would be most useful as a data base in developing it. Some elements are:

- Annual landings by geographic subdivisions, with short- and long-term trends and fluctuations
- Population estimates from trawling surveys, including age structure, age-specific fecundity, and average annual age-specific survival
- Annual larval abundance and distribution estimates from plankton surveys
- Data from routine long-term monitoring of physical/chemical variables over the entire range of the species, but emphasizing spawning/nursery areas
- Descriptions and detailed maps of the levels of principal pollutants (PAH, PCBs, heavy metals) in sediments and water column in important coastal/estuarine habitats of winter flounder
- A history of the nature and degree of pollution in each major estuary important in the life history of winter flounders
- Additional descriptions of the effects of pollutants on developing eggs and larvae, with projections of population impacts at various pollutant levels
- Additional descriptions of the effects of pollutants on juvenile and adult fish, with estimates of population impacts at various pollutant levels
- Documentation of changes in habitats over time

* Dr. F. Thurberg, NMFS, Northeast Fisheries Science Center, Milford, CT, is testing the model developed by Mr. F. Almeida and Dr. M. Fogarty, Northeast Fisheries Science Center, Woods Hole, MA.

Needed, then, is information on:

- Population size, age structure, and age-specific fecundity
- Natural mortality rates
- Fishing mortality rates
- Pollution-induced mortality, especially in early life history stages

Simulation models can then be constructed to predict:

- Population effects of a one-time acute pollution event (oil spill, other)
- Population effects of long-term increasing levels of pollution
- Population effects of long-term decreasing levels of pollution
- Long-term negative genetic effects

DISCUSSION

The case histories of quantitative pollution effects on populations of menhaden, striped bass, and winter flounder presented here are of course only examples. Some information is available for certain other species such as salmon and shad. If this paper were to be written from a European perspective, the long-term research of German workers (Rosenthal, Dethlefsen, von Westernhagen, and their colleagues) examining pollutant effects on larval survival and development would certainly be emphasized. That series of studies began in 1967 and continues to the present time.[51] Among the many significant general findings reported by those investigators are these:

1. Exposure of maturing females to low concentrations of contaminants — especially those which are bioaccumulated — can affect gonad tissue, with effects expressed in the next generation.
2. Life cycle stages most vulnerable to contaminants are maturing females, early embryos, early hatched larvae, and larvae in transition from yolk-sac to feeding.
3. A wide range of morphological, behavioral, and physiological abnormalities in larvae result from exposure to contaminants, in rough proportion to the environmental level of the particular contaminant.
4. Common morphological abnormalities include malformed lower jaw, eye deformities, anomalies in the vertebral column, and reduced size at hatching.
5. Common physiological abnormalities include reduced heart rate, reduced swimming ability, disturbance in equilibrium, and reduced feeding.
6. Early developmental stages showed the highest malformation rates.

Of course the basic question that must be asked is, *"Can defective embryonic development and high embryo mortality due to pollution affect recruitment?"* Some observations relative to this question have been proposed by German investigators.[52] According to their reasoning, total mortality during the embryonic stage of development of marine fish has been estimated to be high — 95 to 99% for species such as Baltic cod and plaice, for example. At this mortality level, decreases in survival rates due to embryo abnormalities, at observed levels from 22 to 33%, would be too small to detect in unexploited populations, but in *overexploited populations* in which spawning stocks have been reduced severely, the added impact of abnormal embryonic development and high embryo mortality could result in reduced recruitment. The investigators also pointed out that, for the North Sea, the highest prevalences of embryonic malformations occurred in highly polluted areas (off the mouths of the Rhine and Elbe rivers and in the vicinity of the dumping zone for titanium dioxide wastes).

From the perspectives of the research scientist and the resource manager, three rather clear responsibilities exist when considering pollution effects on fisheries: one is to demonstrate that coastal/estuarine pollution is really affecting fish and shellfish *stocks,* the second is to define the local or regional stocks (especially shellfish) that *are* affected, and the third is to propose management measures to *mitigate damage* if it exists.

These responsibilities can be satisfied in four stages, which are easy to state but difficult to accomplish: (1) isolate and quantify pollution's effects on resource species — as distinct from effects of natural environmental variations; (2) conduct critical examinations of pollution effects at levels of the individual, the local population, and the species; (3) encourage the identification and quantification of sensitive early-warning indicators of environmental degradation; and (4) attempt to reduce pollutant inputs, where damage to living resources has been or can be demonstrated. Item 4 is particularly difficult to achieve; it points out the pressing need for aquatic scientists to interact with those responsible for managing and regulating terrestrial sources of pollution.

Case histories of the quantitative effects of pollution on populations of menhaden, striped bass, and winter flounder, as presented in this chapter, illustrate a series of generalizations useful in stock management:

1. In assessing pollution impacts, the *entire range* of the species should be considered, and precise information about levels of all contaminants throughout that range should be available. Furthermore, migratory characteristics must be considered, since some species may move rapidly into or out of areas of severe pollution, while others may become semipermanent residents of such degraded areas.
2. Particular attention must be paid to controlling pollution levels in spawning/nursery areas, since most pollution-associated mortality will occur during the first year of life. It is generally accepted by fish population biologists that survival during the first 60 days of life determines the size of the year-class of many estuarine-dependent species; factors affecting early survival (such as pollution) should be of concern and subjects of management action.
3. "...very little impact [from pollution] should be tolerated at low stock size, because it would prevent recovery to a...maximum sustained yield."[53]
4. Simulation models can supply useful information, but the degree of reliability of such models depends on the extent of the data base employed.

5. Marine/anadromous fish populations are characterized by aperiodic dominant year-classes, which may form the basis for a fishery for many years. (An example would be the 1970 striped bass year-class in Chesapeake Bay.) Existence of these dominant year-classes, and examination of factors responsible for their production, can lead to important insights about environmental influences on abundance.

Previous sections of this chapter included examinations of case histories that indicated points in the life cycles of fish where pollutant stressors can exert lethal and sublethal influences. From experimental studies, it seems obvious that with all these potential impact points throughout life cycles, populations in contaminated waters should dwindle and disappear; yet from experience on the east coast of North America, this has not happened (at least not yet). Only in severely degraded local waters, which are a small part of the total range of most fish species, have there been localized disappearances; even in those areas, other species that might be expected to be affected are still present and are in some instances abundant. This is true particularly of a number of coastal/estuarine-dependent fish species, many of which spend much of their life cycles in waters that are to some extent contaminated. It may also be true for saltatory species that can move into or through degraded zones, although any possible impacts of transient exposures to contaminants are more difficult to recognize and much more difficult to quantify.

A statistical analysis of the fishery resources of the U.S., published in 1974,[54] discussed catches of all important marine fish and shellfish, on a species-by-species basis, for the years 1950 to 1970. Included were the major Atlantic species usually considered estuarine-dependent or estuarine and near-shore inhabitants, species that might be expected to demonstrate some effects of increasing estuarine and coastal pollution. With few exceptions, according to that publication, catches of those species were relatively steady or showed some increases in the 20-year period. One of the summarizing statements in the report is of particular note: "In general there are no good fishing effort data available for those estuarine species whose catches have remained constant, so it is possible that maintenance of catch levels is due to constantly increasing fishing effort. *Nonetheless, the evidence from catch records of a substantial number of exploited estuarine species in United States waters indicates that pollution and damage to estuaries have not yet shown any measurable overall effect on the part of the marine resource which might be expected to show the first effects*" [italics added]. Admittedly, the report covered only the period from 1950 to 1970, but for at least some of the species considered, the situation has not changed dramatically since that time (with the possible exception of striped bass — but even here, overfishing seems to have been the principal cause of the most recent (1973 to 1986) decline in abundance).

Two major recent attempts have been made to assess pollution impacts on fish stocks off portions of the U.S. coasts — one by Prager and MacCall[55] examining data from the southern California bight and the other by Summers et al.[56] using historical data for a number of fish species from five estuaries of the Northeastern states. Using two modeling methods — a biomass-based model and a recruitment model — Prager and MacCall looked for significant effects of climatic conditions and contaminant loadings on the spawning success of three Pacific species: northern anchovy, *Engraulis mordax,* Pacific sardine, *Sardinops sagax,* and chub mackerel, *Scomber japonicus*. Using the models, no climate or contaminant influences on the spawning success of northern anchovies were detected, but the spawning success of chub mackerel seemed to be strongly influenced by climatic variability, and (with the recruitment model) the spawning success of Pacific sardines was strongly negatively correlated with contaminant loadings — being consistent with the hypothesis that the stock, which had been overfished and had collapsed in the 1950s, was stressed beyond its limits by poor larval survival caused by ambient contaminant concentrations. The study of fish populations of five Atlantic coast estuaries by Summers et al.,[56] using a biostatistical modeling approach based on catch statistics, disclosed consistent patterns of pollution effects among similar species across different estuaries.

Another analysis — this one of recent changes in abundance of North Sea fish stocks — failed to find any effect from pollution. Tiews[57] examined the abundance trends of 25 species for 35 years (1954 to 1988) from trawl catches on the German North Sea coast. He found that some species had declined, some had increased, and others had fluctuated irregularly, but that no consistent long-term decline of commercial species could be attributed to deteriorating habitat conditions. That noted investigator had earlier (1983) expressed suspicion that, at least for some species, declines might be due to pollution effects, but his latest analysis concluded with the comment that "...this study shows that the majority of species studied seems to be able to tolerate the present status of environmental deterioration of the ecosystem. Furthermore, the study does not indicate any fundamental impairment of the fishery biological situation of the area during the last seven years in comparison to the preceding period from 1954 to 1981). Six species have even substantially increased in abundance." Tiews cautioned, however, that 35 years may be too short a period to reveal natural population fluctuations, so that a very conservative interpretation of the data is necessary.

A well-known quantitative scientist from the U.S., C.P. Goodyear,[58] has described the dimensions of the problem of interpreting population responses to direct (lethal) and indirect (sublethal) levels of contaminants very succinctly:

- For cases where contaminants are directly lethal, two types of information are required: "...the timing and extent of the excess mortality must be determined." "...the nature, timing, and intensity of the density-dependent processes that regulate the size of the population must be understood" (this, as Goodyear, pointed out, is "a key, largely unsolved problem in fishery research and management in general").

- For cases where indirect (sublethal) levels of contaminants exist, "The interpretation of the population response to the indirect effects of the contaminants on the species of interest requires quantification of the response in terms of a change in survival probability or a change in reproductive rates, and knowledge of the density-dependent processes that control population size." (Goodyear then pointed out that "none of the required information is particularly amenable to measurement"!)

Experimental demonstrations of the negative effects of chemical contaminants on the survival and well-being of marine fish and shellfish are abundant, but the variability that results from the influences of natural factors is so great and so incompletely understood that experimental findings cannot usually be applied directly to assessments of exploited populations — despite repeated attempts to do so. Some extrapolations from single-contaminant experimental studies can be useful, however.

One recent paper[59] described calculations of year-class reductions in cod, herring, and saithe populations of the Norwegian coast that might result from a major oil spill. Detailed field studies of seasonal herring larval distribution and concurrent laboratory studies of effects of oil on larvae, when presented in a worst case scenario, enabled calculation of what turned out to be a low percentage of potential reduction in recruitment. Field experimental studies, in which fish are exposed *in situ* to contaminated waters, have also produced meaningful results when combined with adequate chemical analyses of environmental and tissue samples.

The appearance of new reports of sublethal effects of contaminants on marine animals has been aptly described as of "avalanche" proportions.[60] In a thought-provoking discussion of options in environmental management and the deployment of future research efforts concerned with the population effects of pollution, the author of that report, made the point in 1980 that "The more subtle threats of chronic pollution well away from the hot-spots, and which because of their subtlety have always been seen as potentially the most dangerous, do not appear to have developed and produced effects on the scale that was feared a decade ago." The author then asked difficult questions: "If after all the recent, intensive effort there is still difficulty in finding chronic effects upon communities does this not suggest that such effects are negligible? That while the initial concern over chronic effects was justified is it not now time to acknowledge that it is only acute pollution that matters?" Most of us in pollution research would detect a strong odor of heresy in this superficially logical line of thought, probably rejecting it out-of-hand, since it tends to overestimate human ability to sort out the effects of natural versus man-induced population changes, and it tends toward too easy acceptance of the significance of acute pollution events in reducing overall abundance of marine populations. However, we would have to admit that our arguments would not have a strong base in adequately demonstrated population impacts. We might also admit that, just as with certain terrestrial species, a threshold may be reached where contaminants in the marine environment could achieve a level, or operate over sufficient time, to greatly affect population abundance.

Assessment of the quantitative effects of pollution and other human-induced environmental perturbations on fish populations depends on an understanding of the stock-recruitment relationships of fish populations of concern, and on knowledge of the density-dependent and density-independent factors influencing those relationships. Of particular concern is mortality imposed by toxic levels of pollutants on eggs, larvae, and juveniles of economically important fish species. However, models predicting the effects of human intrusions (other than fishing) must be used conservatively, especially when impacts on early life stages are included. The concern has been expressed with precision as follows[61]: "...the potential for error in numerical predictions of the effect of proposed levels of increased mortality on pre-recruit stages is large, while the biologically acceptable range of error is small," and further that "Until a better understanding is achieved of the interacting roles of density-dependent and density-independent factors in regulating population size and stability, and until a much better data base is available for the majority of the stocks subject to mortality from industrial activity, precise numerical predictions of the impact of this incremental mortality on adult stocks should be interpreted with great caution."

Assuming the correctness of the generally held principle that recruitment variability is a consequence of events that occur in egg, larval, and postlarval stages, it is logical to focus pollution studies more directly on reproductive success of populations at risk. This has been done in a number of studies, but one disturbing conclusion that can be reached after examining some of the accumulated literature is that *investigators rarely if ever carry their findings to the point of actually estimating quantitative effects on fish populations and species*. Let's consider a few examples. German researchers[62] examined the contaminant content of ovaries of Baltic flounder, *Platichthys flesus,* and established a threshold level of 0.12 ppm PCBs beyond which reproductive impairment would occur. These authors found that 8.5% of the sample exceeded that level (range, 0.05 to 3.17), and they reported that viable hatch was lower than 15% if gonad PCBs exceeded 0.25 ppm. They did *not* take the final quantitative steps. The Baltic flounder catches are known; the extent of PCB contamination in the Baltic is known. Why not make a rough estimate from these data of the possible impact of PCBs contamination on recruitment of Baltic flounders? (And then even make an economic evaluation as well.) As another example, American scientists[63] found reproductive impairment and high levels of ovarian DDT and PCB in white croakers, *Genyonemus lineatus,* from a contaminated site near Los Angeles. Again, white croaker catch statistics are available, and environmental levels of chlorinated hydrocarbons on the California coast have been examined extensively. Why not take another step, armed with the data, and provide a rough estimate of the possible impact on recruitment?

Only once, to my knowledge, has even the penultimate step — estimating the effects of pollution on recruitment — been taken. This was a report on developmental defects in pelagic fish embryos from the Western Baltic.[64] Plankton net catches of cod, plaice, and flounder eggs and larvae were examined for mortality and abnormalities. Prevalences of defective embryos and abnormal larvae in various locations in the western Baltic were determined, and the effects on survival of larvae of two year-classes (1983 and 1984) were estimated. Decreases (18 to 44%) were considered too small in terms of biological significance to cause a detectable impact on recruitment, although the authors pointed out that even a small impact on larval production could, in the case of overexploited Baltic stocks, lead eventually to reduced recruitment. It is puzzling, considering the availability of excellent fishery statistics and environmental information for the Baltic, some of it extending back for more than a century, why the ultimate step — of estimating total possible population effects of existing pollution levels — was not taken, since it is only at this stage that pollution information would become meaningful to resource managers.

Part of the problem may be that those professionals interested in problems created by coastal/estuarine pollution do not communicate effectively with population dynamics specialists, so full exploitation of available data does not happen. Only in rare instances, such as the mid-Atlantic striped bass program discussed earlier, when simulations based on good resource and environmental data sets are attempted, is some measure of integration achieved. The assessment scientist is often loath to use the minimalist approach necessary with the environmental scientist's data.

To do an effective job of quantifying pollution impacts on fish stocks, a large interdisciplinary research and monitoring program is required. The principal ingredients are shown in Figure 34. A triumvirate of population assessments, environmental assessments, and experimental studies constitutes the basic information source for the required modeling effort. Such a program must be long-term as well as geographically broad. Until now, only the mid-Atlantic striped bass program has approached the level of research commitment required, and even this major effort must face an annual struggle for adequate funding, in spite of its considerable importance and high levels of lobbying.

CONCLUSIONS

More than a decade ago, in a major symposium titled "Protection of Life in the Sea," held at the Biologische Anstalt Helgoland, a series of conclusions was proposed about pollution effects on fisheries — conclusions that seem still relevant today:[65]

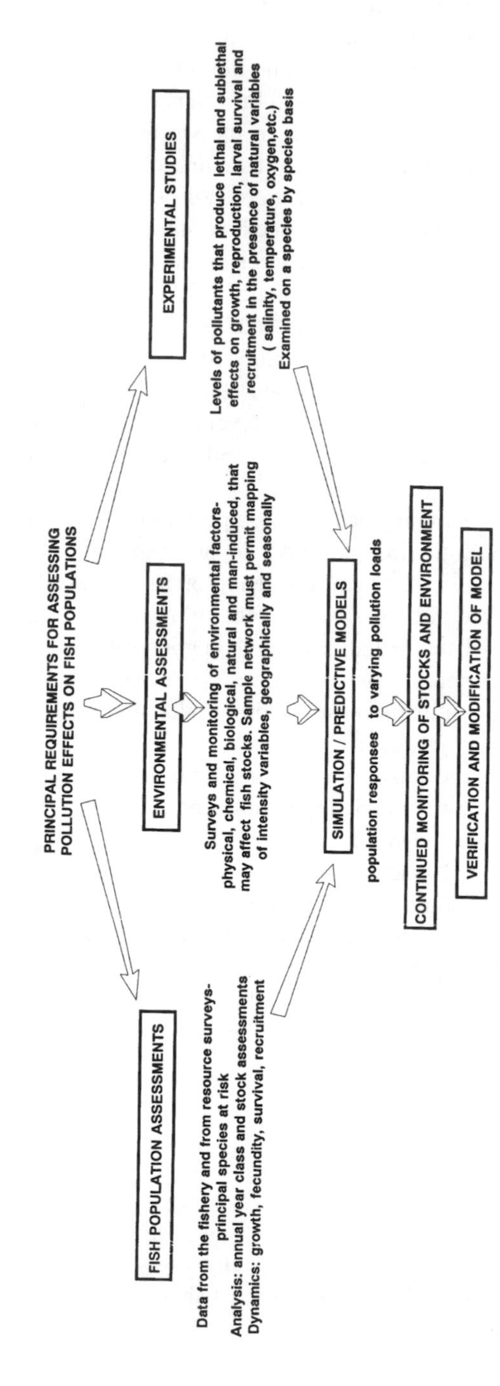

FIGURE 34. Principal requirements for assessing the effects of pollution on fish stocks.

Pollution effects on fisheries have received some scrutiny in recent decades, and information is accumulating, but is still insufficient to be very useful in resource management decisions — except as they involve local areas. Evidence exists for localized effects of pollutant stress on fisheries, but *as yet there is little specific evidence for widespread damage to major fisheries resource populations resulting from coastal/estuarine pollution*. This may well be because we are unable to separate clearly the effects of pollutant stress from effects of the many other forms of environmental stresses to which marine populations are subject. Other factors, such as shifts in geographic distribution of fish populations, changes in productive ecosystems, or overfishing, may cause pronounced changes in fisheries — changes which could obscure any effects of localized habitat degradation. *It seems, with the evidence presently available, that factors other than pollution are overriding in determining fish abundance, but we lack sufficient quantitative data to make positive statements about cause and effect relationships of abundance and pollution.*

It may be, of course, that coastal/estuarine pollution is exerting some overall influence on certain resource species, but that this influence may be masked by increased fishing effort, or by favorable changes in other environmental factors which create a positive effect on abundance, outweighing any negative effects of pollutants. Many experimental studies, particularly more recent ones concerned with long-term exposure of fish and shellfish to low levels of contaminants, suggest that some long-term effects on abundance should be felt, but our statistics, our monitoring, and our population assessments are not yet adequate to detect them.

Effective long-term monitoring of stocks and environment must be the basis of any attempt to isolate and identify pollution effects. *A continuous integrated effort in stock assessment, environmental assessment, and experimental studies will be required to understand the role of all environmental stressors — natural and man-induced — in determining abundance of resource populations*. However, we do not yet have the principal pieces of the puzzle in place, so in the absence of full understanding of the phenomena involved, management decisions affecting coastal/estuarine pollution must be made on the basis of "best available scientific information," just as decisions about allowable resource exploitation are made. In both types of decision processes, a conservative action provides a lower risk of damage and loss than does a more extreme action.

Conservatism can be especially significant when decisions are made that might permit pollution to continue or increase, since long-term effects of existing levels on abundance of resource populations are largely undetermined. In addition to advocating conservatism, we must persist in attempts to quantify the effects of pollution, and to determine the precise pathways through which fishery resources are affected.

Freely admitting to a conservative mind-set (and a mild infatuation with my own prose), there is little about these conclusions that I would change today — more than a decade after their original publication. Research in the intervening years has added *substantially* but only *incrementally* and not *conceptually* to our understanding of pollution effects on resource species; some of that information has been summarized in the case histories just presented.

The major problem — being able to distinguish adequately the effects of pollution from all the other influences on marine fish population abundance — still confronts us, even though the boundaries of our knowledge have expanded. Recent advances include the creation and implementation of new pollution monitoring and assessment programs, developments in simulation modeling, annual lengthening of critical resource and environmental data sets, and findings from field and laboratory experimental studies.

Some pessimists feel that we will never have adequate data to distinguish clearly the quantitative effects of pollution on fish stocks; I think the accumulation of analyses and insights relevant to the problem forecasts a brighter future than that.

REFERENCES

1. **Cole, H. A.** 1972. North Sea pollution, pp. 3–9. in Ruivo, M. (Ed.), *Marine Pollution and Sea Life*. Fishing News Ltd., London.
2. **Cross, F. A., D. S. Peters, and W. E. Schaaf.** 1985. Implications of waste disposal in coastal waters on fish populations, pp. 383–399. in Cardwell, R. D., R. Purdy, and R. C. Bahner (Eds.), *Aquatic Toxicology and Hazard Assessment: Seventh Symposium,* ASTM-STP 854. American Society for Testing and Materials, Philadelphia, PA.
3. **Lee, A.** 1978. Effects of man on the fish resources of the North Sea. *Rapp. P.-V. Réun. Cons. Int. Explor. Mer* 173: 231–240.
4. **Dethlefsen, V. and K. Tiews.** 1985. Review of the effects of pollution on marine fish life and fisheries in the North Sea. *Z. Angew. Ichthyol.* 1: 97–118.
5. **Hempel, G.** 1978. North Sea fisheries and fish stocks — a review of recent changes. *Rapp. P.-V. Reun. Cons. Int. Explor. Mer* 173: 145–167.
6. **Waldichuk, M.** 1979. Review of the problems, pp. 399–424. in Cole, H.A. (Ed.), *The Assessment of Sublethal Effects of Pollutants in the Sea. Philos. Trans. R. Soc. Lond.* B286.
7. **Cushing, D. H.** 1979. The monitoring of biological effects: the separation of natural changes from those induced by pollution, pp. 597–609. in Cole, H.A. (Ed.), *The Assessment of Sublethal Effects of Pollutants in the Sea. Philos. Trans. R. Soc. Lond.* B286.
8. **Jones, R.** 1982. Population fluctuations and recruitment in marine populations. *Philos. Trans. R. Soc. Lond.* B297: 353–368.
9. **Segar, D. A. and E. Stamman.** 1986. Monitoring in support of estuaries pollution management needs, pp. 874–877. in *Proceedings of Oceans '86,* Vol. 3. Mar. Technol. Soc., Washington, D.C.
10. **Dethlefsen, V.** 1986. Marine pollution mismanagement: towards the precautionary concept. *Mar. Pollut. Bull.* 17: 54–57.
11. **Vaughan, D. S., P. Kanciruk, and J. E. Breck.** 1982a. Research needs to assess population-level effects of multiple stresses on fish and shellfish. Oak Ridge Natl. Lab., ORNL/TM-8375, Oak Ridge, TN; **Vaughan, D. S., R. M. Yoshiyama, J. E. Breck, and D. L. DeAngelis.**1982b. Review and analysis of existing modeling approaches for assessing population-level effects of multiple stresses on fish and shellfish. Oak Ridge Natl. Lab., ORNL/TM-8342,

Oak Ridge, TN; **Kanciruk, P., J. E. Breck, and D. S. Vaughan.** 1982. Population-level effects of multiple stresses on fish and shellfish. Oak Ridge Natl. Lab., ORNL/TM-8317, Oak Ridge, TN.

12. **Schaaf, W. E., D. S. Peters, D. S. Vaughan, L. Coston-Clements, and C. W. Krouse.** 1987. Fish population responses to chronic and acute pollution: the influence of life history strategies. *Estuaries* 10: 267–275.
13. **Vaughan, D. S., J. V. Merriner, and W. E. Schaaf.** 1986. Detectability of a reduction in a single year class of a fish population. *J. Elisha Mitchell Sci. Soc.* 102: 122–128.
14. **Van Winkle, W. (Ed.).** 1977. *Proceedings of the Conference on Assessing the Effects of Power-Plant-Induced Mortality on Fish Populations.* Pergamon Press, New York. 380 pp.; **Van Winkle, W., D. S. Vaughan, L. W. Barnthouse, and B. L. Kirk.** 1981. An analysis of the ability to detect reduction in year-class strength of the Hudson River white perch *(Morone americana)* population. *Can. J. Fish. Aquat. Sci.* 38: 627–632; **Vaughan, D. S. and W. Van Winkle.** 1982. Corrected analysis of the ability to detect reductions in year-class strength of the Hudson River white perch *(Morone americana)* populations. *Can. J. Fish. Aquat. Sci.* 39: 782–785.
15. **Swartzman, G., R. Deriso, and C. Cowan.** 1977. Comparison of simulation models used in assessing the effects of power-plant-induced mortality on fish populations, pp. 333–361. in Van Winkle, W. (Ed.), *Proceedings of the Conference on Assessing the Effects of Power-Plant-Induced Mortality on Fish Populations.* Pergamon Press, New York.
16. **Merriner, J. V.** 1976. Differences in management of marine recreational fisheries, pp. 123–131. in Clepper, H. (Ed.), *Marine Recreational Fisheries.* Sport Fishing Institute, Washington, D.C.; **Goodyear, C. P.** 1978. Management problems of migratory stocks of striped bass, pp. 75–84. in Clepper, H. (Ed.), *Marine Recreational Fisheries 3.* Sport Fishing Institute, Washington, D.C.; **Goodyear, C. P.** 1980. Oscillatory behavior of a striped bass population model controlled by a Ricker function. *Trans. Am. Fish. Soc.* 109: 511–516; **Goodyear, C. P.** 1984a. Analysis of potential yield per recruit for striped bass produced in the Chesapeake Bay. *N. Am. J. Fish. Manage.* 4: 488–496; **Goodyear, C. P.** 1984b. Measuring effects of contaminant stress on fish populations, pp. 414–424. in *Aquatic Toxicology: Sixth Symposium.* Spec. Tech. Publ. No. 802, American Society for Testing and Materials, Philadelphia, PA; **Levin, S. A. and C. P. Goodyear.** 1980. Analysis of an age-structured fishery model. *J. Math. Biol.* 9: 245–274; **Florence, B. M.** 1980. Harvest of the northeastern coastal striped bass stocks produced in the Chesapeake Bay, pp. 29–44. in Clepper, H. (Ed.), *Marine Recreational Fisheries 5.* Sport Fishing Institute, Washington, D.C.; **Mehrle, P. M., D. Buckler, S. E. Finger, and L. Ludke.** 1984. Impact of contaminants on striped bass. U.S. Fish and Wildlife Service, Columbia Natl. Fish. Res. Lab., Interim Rep., Columbia, MO. 28 pp.; **Boreman, J. and H. M. Austin.** 1985. Production and harvest of anadromous striped bass stocks along the Atlantic coast. *Trans. Am. Fish. Soc.* 114: 3–7; **Goodyear, C. P., J. E. Cohen, and S. W. Christensen.** 1985. Maryland striped bass: recruitment declining below replacement. *Trans. Am. Fish. Soc.* 114: 146–151.
17. **Munro, A. L. S., A. H. McVicar, and R. Jones.** 1983. The epidemiology of infectious disease in commercially important wild marine fish. *Rapp. P.-V. Réun. Cons. Int. Explor. Mer* 182: 21–32..

18. **Mehrle, P. M., T. A. Haines, S. Hamilton, J. L. Ludke, F. L. Mayer, and M. A. Ribick.** 1982. Relationship between body contaminants and bone development in east-coast striped bass. *Trans. Am. Fish. Soc.* 111: 231–241; **Pizza, J. C. and J. M. O'Connor.** 1983. PCB dynamics in Hudson River striped bass. II. Accumulation from dietary sources. Aquat. Toxicol. 3: 313–327; **Hall, L. W., Jr., L. O. Horseman, and S. Zeger.** 1984. Effects of organic and inorganic chemical contaminants on fertilization, hatching success, and prolarval survival of striped bass. *Arch. Environ. Contam. Toxicol.* 13: 723–729; **Hall, L. W., Jr., A. E. Pinkney, R. L. Herman, and S. E. Finger.** 1987. Survival of striped bass larvae and yearlings in relation to contaminants and water quality in the upper Chesapeake Bay. *Arch. Environ. Contam. Toxicol.* 16: 391–400.
19. **Hall, L. W., Jr., A. E. Pinkney, L. O. Horseman, and S. E. Finger.** 1985. Mortality of striped bass larvae in relation to contaminants and water quality conditions in a Chesapeake Bay tributary. *Trans. Am. Fish. Soc.* 114: 861–868.
20. **Mehrle, P. M., T. A. Haines, S. Hamilton, J. L. Ludke, F. L. Mayer, and M. A. Ribick.** 1982. Relationship between body contaminants and bone development in east coast striped bass. *Trans. Am. Fish. Soc.* 111: 231–241.
21. **Hunt, E. and J. Linn.** 1969. Fish kills by pesticides, pp. 44–59. in Gillette, J. W. (Ed.), *Proceedings of the Symposium on the Biological Impact of Pesticides in the Environment.* Oregon State University, Corvallis.
22. **Westin, D. T., C. E. Olney, and B. A. Rogers.** 1983. Effects of parental and dietary PCBs on survival, growth, and body burdens of larval striped bass. *Bull. Environ. Contam. Toxicol.* 30: 50–57; **Westin, D. T., C. E. Olney, and B. A. Rogers.** 1985. Effects of parental and dietary organochlorines on survival and body burdens of striped bass larvae. *Trans. Am. Fish. Soc.* 114: 125–136.
23. **Mehrle, P. M., T. A. Haines, S. Hamilton, J. L. Ludke, F. L. Mayer and M. A. Ribick.** 1982. Relationship between body contaminants and bone development in east coast striped bass. *Trans. Am. Fish Soc.* 111: 231–234.
24. **Goodyear, C. P.** 1985b. Toxic materials, fishing, and environmental variation: simulated effects on striped bass population trends. *Trans. Am. Fish. Soc.* 114: 107–113.
25. **Saila, S. B.** 1962. The contribution of estuaries to the offshore winter flounder fishery in Rhode Island. *Proc. Gulf Caribb. Fish. Inst.* 1961: 95–105; **Berry, R. J., S. B. Saila, and D. B. Horton.** 1965. Growth studies of winter flounder, *Pseudopleuronectes americanus* (Walbaum), in Rhode Island. *Trans. Am. Fish. Soc.* 94: 259–264.
26. **Bigelow, H. B. and W. C. Schroeder.** 1953. Fishes of the Gulf of Maine. *U.S. Fish and Wildlife Service, Fish. Bull.* 53, 577 pp.; **Hughes, J. B., D. A. Nelson, D. M. Perry, J. E. Miller, G. R. Sennefelder, and J. J. Periera.** 1986. Reproductive success of the winter flounder *(Pseudopleuronectes americanus)* in Long Island Sound. *Int. Counc. Explor. Sea,* Doc. C.M.1986/E:10, 11 pp.
27. **Topp, R. W.** 1967. Biometry and related aspects of young winter flounder, *Pseudopleuronectes americanus* (Walbaum), in the Weweantic River estuary. M.S. Thesis, University of Massachusetts. 65 pp.; **Smith, R. M. and C. F. Cole.** 1970. Chlorinated hydrocarbon insecticide residues in winter flounder, *Pseudopleuronectes americanus,* from the Weweantic River estuary, Massachusetts. *J. Fish. Res. Board Can.* 27: 2374–2380.

28. **Burdick, G. E., E. J. Harris, H. J. Dean, T. M. Walker, J. Skea, and D. Colby.** 1964. The accumulation of DDT in lake trout and the effect on reproduction. *Trans. Am. Fish. Soc.* 93: 127–136; **Allison, D., B. J. Kallman, O.B. Cope, and C. van Valin.** 1964. Some chronic effects of DDT on cutthroat trout. *U.S. Fish and Wildlife Service, Res. Rep.* 64, 30 pp.; **Johnson, H. E. and C. Pecor.** 1969. Coho salmon mortality and DDT in Lake Michigan. *Trans. 34th N. Am. Wildl. Natur. Resour. Conf.,* pp. 159–166.
29. **Smith, R. M. and C. F. Cole.** 1973. Effects of egg concentrations of DDT and dieldrin on development in winter flounder *(Pseudopleuronectes americanus). J. Fish. Res. Board Can.* 30: 1894–1898.
30. **Longwell, A. C., D. Perry, J. B. Hughes, and A. Herbert.** 1983. Frequencies of micronuclei in mature and immature erythrocytes of fish as an estimate of chromosome mutation rates — results of field surveys on windowpane flounder, winter flounder and Atlantic mackerel. *Int. Counc. Explor. Sea,* Doc. C.M.1983/E:55.
31. **Black, D. E., D. K. Phelps, and R. L. Lapan.** 1988. The effect of inherited contamination on egg and larval winter flounder, *Pseudopleuronectes americanus. Mar. Environ. Res.* 25: 45–62.
32. **Kühnhold, W. W., D. Everich, J. J. Stegeman, J. Lake, and R. E. Wolke.** 1978. Effects of low levels of hydrocarbons on embryonic, larval and adult winter flounder *(Pseudopleuronectes americanus),* pp. 677–711. in *Proceedings of the Conference on Assessment of Ecological Impacts of Oil Spills,* Keystone, Colorado. *American Institute of Biological Science,* Washington, D.C.
33. **McIntyre, A. D.** 1982. Oil pollution and fisheries, pp. 401–411, in Clark, R. B. (Ed.) *The Long Term Effects of Oil Pollution Marine on Populations, Communities, and Ecosystems.* Phil. Trans. R. Soc. Lond. B297.; **Laurence, G. C.** 1975. Laboratory growth and metabolism of the winter flounder *Pseudopleuronectes americanus* from hatching through metamorphosis at three temperatures. *Mar. Biol.* 32: 223–229; **Laurence, G. C.** 1977. A bioenergetic model for the analysis of feeding and survival potential of winter flounder, *Pseudopleuronectes americanus,* larvae during the period from hatching to metamorphosis. *Fish. Bull.* 75(3): 529–546.
34. **Nelson, D. A., J. E. Miller, D. Rusanowsky, R. A. Greig, G. R. Sennefelder, R. Mercaldo-Allen, C. Kuropat, E. Gould, F. P. Thurberg, and A. Calabrese.** 1991. Comparative reproductive success of winter flounder in Long Island Sound: a 3-year study (biology, biochemistry, and chemistry). *Estuaries* 14: 318–331; **Perry, D. M., J. B. Hughes, and A. T. Hebert.** 1991. Sublethal abnormalities in embryos of winter flounder, *Pseudopleuronectes americanus,* from Long Island Sound. *Estuaries* 14: 306–317.
35. **Butler, P. A., R. Childress, and A. J. Wilson.** 1972. The association of DDT residues with losses in marine productivity, pp. 262–266. in Ruivo, M. (Ed.), *Marine Pollution and Sea Life.* Fishing News Ltd., London.
36. **Westernhagen, H. von, H. Rosenthal, V. Dethlefsen, W. Ernst, U. Harms, and P.-D. Hansen.** 1981. Bioaccumulating substances and reproductive success in Baltic flounder *Platichthys flesus. Aquat. Toxicol.* 1: 85–99; **Westernhagen, H. von, P. Cameron, V. Dethlefsen, and D. Janssen.** 1989. Chlorinated hydrocarbons in North Sea whiting (*Merlangius merlangus* (L.)) and effects on reproduction. *Helgol. Meeresunters.* 43: 45–60.

37. **Hansen, P.-D., H. von Westernhagen, and H. Rosenthal.** 1985. Chlorinated hydrocarbons and hatching success in Baltic herring spring spawners. *Mar. Environ. Res.* 15: 59–76.
38. **Baker, J. T. P.** 1969. Histological and electron microscopical observations on copper poisoning in the winter flounder *(Pseudopleuronectes americanus). J. Fish. Res. Board Can.* 26: 2785–2793.
39. **Calabrese, A., F. P. Thurberg, M. A. Dawson, and D. R. Wenzloff.** 1975. Sublethal physiological stress induced by cadmium and mercury in the winter flounder, *Pseudopleuronectes americanus,* pp. 15–21. in Koeman, J. H. and J.J.T.W.A. Strik (Eds.), *Sublethal Effects of Toxic Chemicals on Aquatic* Animals. Elsevier, Amsterdam; **Pritchard, J. B., A. M. Guarino, and W. B. Kinter.** 1973. Distribution, metabolism, and excretion of DDT and mirex by a marine teleost, the winter flounder. *Environ. Health Perspect.* 4: 45–54.
40. **Zdanowicz, V. S., D. F. Gadbois, and M. W. Newman.** 1986. Levels of organic and inorganic contaminants in sediments and fish tissues and prevalences of pathological disorders in winter flounder from estuaries of the northeast United States, 1984. in *IEEE Oceans '86 Conference Proceedings,* pp. 578–585. Washington, D.C.
41. **Fletcher, G. L., J. W. Kiceniuk, and U. P. Williams.** 1981. Effects of oiled sediments on mortality,feeding and growth of winter flounder *Pseudopleuronectes americanus. Mar. Ecol. Prog. Ser.* 4: 91–96.
42. **Murchelano, R. A. and R. E. Wolke.** 1985. Epizootic carcinoma in the winter flounder, *Pseudopleuronectes americanus. Science* 228: 587–589.
43. **Malins, D. C., B. B. McCain, D. W. Brown, S.-L. Chan, M. S. Myers, J. T. Landahl, P. G. Prohaska, A. J. Friedman, L. D. Rhodes, D. G. Burrows, W. D. Gronlund, and H. O. Hodgins.** 1984. Chemical pollutants in sediments and diseases of bottom-dwelling fish in Puget Sound, Washington. *Environ. Sci. Technol.* 13: 705–713; **Malins, D. C., B. B. McCain, J. T. Landahl, M. S. Myers, M. M. Krahn, D. W. Brown, S.-L. Chan, and W. T. Roubal.** 1988. Neoplastic and other diseases in fish in relation to toxic chemicals: an overview. *Aquat. Toxicol.* 11: 43–67.
44. **McMahon, G., L. J. Huber, J. J. Stegeman, and G. N. Wogan.** 1988. Identification of a C-Ki-ras oncogene in a neoplasm isolated from winter flounder. *Mar. Environ. Res.* 24: 345–350.
45. **Murchelano, R. A.** 1988. Fish as sentinels of environmental health. U.S. Department of Commerce, NOAA Tech. Memo. NMFS-F/NEC-61, 16 pp.
46. **Payne, J. F. and L. F. Fancey.** 1989. Effect of polycyclic aromatic hydrocarbons on immune responses in fish: change in melanomacrophage centers in flounder *(Pseudopleuronectes americanus)* exposed to hydrocarbon-contaminated sediments. *Mar. Environ. Res.* 28: 431–435.
47. **Wolke, R. E., C. J. George, and V. S. Blazer.** 1984. Pigmented macrophage accumulations (MMC; PMB): possible monitors of fish health, pp. 93–97. in Hargis, W. (Ed.), *USA-USSR Symposium on Pathogens and Parasites of the World Oceans,* Leningrad, October 1981. U.S. Department of Commerce, NOAA, Natl. *Mar. Fish. Serv. Tech. Rep.* 25; **Wolke, R. E., R. A. Murchelano, C. D. Dickstein, and C. J. George.** 1985. Preliminary evaluation of the use of macrophage aggregates (MA) as fish health monitors. *Bull. Environ. Contam. Toxicol.* 35: 222–227.

48. **Peters, N., A. Köhler, and H. Kranz.** 1987. Liver pathology in fishes from the lower Elbe as a consequence of pollution. *Dis. Aquat. Org.* 2:87–97.
49. **Khan, R. A.** 1987. Effects of chronic exposure to petroleum hydrocarbons on two species of marine fish infected with a hemoprotozoan, *Trypanosoma murmanensis*. *Can. J. Zool.* 65: 2703–2709.
50. **Haedrich, R. L. and S. O. Haedrich.** 1974. A seasonal survey of the fishes of the Mystic River, a polluted estuary in downtown Boston, Massachusetts. *Estuar. Coast. Mar. Sci.* 2: 59–73.
51. **Kinne, O. and H. Rosenthal.** 1967. Effects of sulfuric water pollutants on fertilization, embryonic development and larvae of the herring, *Clupea harengus*. *Mar. Biol.* (Berl.) 1: 65–83; **Rosenthal, H., M. McInerney-Northcott, C. J. Musial, J. F. Uthe, and J. D. Castell.** 1986. Viable hatch and organochlorine contaminant levels on gonads of fall spawning Atlantic herring from Grand Manan, Bay of Fundy, Canada. *Int. Counc. Explor. Sea,* Doc. C.M./E:26; **Dethlefsen, V., P. Cameron, H. von Westernhagen, and D. Janssen.** 1987. Morphologische und chromosomale Untersuchungen an Fischembryonen der südlichen Nordsee in Zusammenhang mit der Organochlorkontamination der Elterntiere. *Veröff. Inst. Küsten Binnenfisch.* 97: 1–57; **Dethlefsen, V.** 1988. Status report on aquatic pollution problems in Europe. *Aquat. Toxicol.* 11: 259–286.
52. **Westernhagen, H. von, V. Dethlefsen, P. Cameron, J. Berg, and G. Fürstenberg.** 1988. Developmental defects in pelagic fish embryos from the western Baltic. *Helgol. Meeresunters.* 42: 13–36.
53. **Cushing, D. H.** 1979. The monitoring of biological effects: the separation of natural changes from those induced by pollution, pp. 597–609. in Cole, H.A. (Ed.), *The Assessment of Sublethal Effects of Pollutants in the Sea. Philos. Trans. R. Soc. Lond.* B286.
54. **Wise, J. P. (Ed.).** 1974. The United States Marine Fishery Resource. U.S. Department of Commerce, NOAA-NMFS, MARMAP 1, 379 pp.
55. **Prager, M. H. and A. D. MacCall.** 1993. Detection of contaminant and climate effects on spawning success of three pelagic fish stocks off southern California: northern anchovy *Engraulis mordax,* Pacific sardine *Sardinops sagax,* and chub mackerel *Scomber japonicus*. *Fish. Bull.* 91:310–327.
56. **Summers, J. K., T. T. Polgar, K. A. Rose, R. A. Cummins, R. N. Ross, and D. G. Heimbuch.** 1986. Assessment of the relationships among hydrographic conditions, macropollution histories, and fish and shellfish stocks in major northeastern estuaries. Martin Marietta Environ. Systems, Contract Rep. NA83-AA-D-00059.
57. **Tiews, K.** 1983. On the changes of fish and crustacean stocks in the German North Sea coast during the years 1954–1981 and the hypothetical role of pollution as a causative factor. *Int. Counc. Explor. Sea,* Doc. C.M.1983/E:16, 18 pp.; **Tiews, K.** 1989. 35 years' abundance trends (1954–1988) of 25 fish and crustacean stocks on the German North Sea coast. *Int. Counc. Explor. Sea,* Doc. C.M.1989/E:28, 11 pp.
58. **Goodyear, C. P.** 1985. Relationship between reported commercial landings and abundance of young striped bass in Chesapeake Bay, Maryland. *Trans. Am. Fish. Soc.* 114: 92–96.

59. **Foyn, L. and B. Serigstad.** 1989. How can a potential oil pollution affect the recruitment to fish stocks? *Int. Counc. Explor. Sea,* Doc. C.M.1989/Minisymp. No. 5, 23 pp.
60. **Lewis, J. R.** 1980. Options and problems in environmental management and evaluation. *Helgol. Meeresunters*. 33: 452–466.
61. **Leggett, W. C.** 1977. Density dependence, density independence, and recruitment in the American shad *(Alosa sapidissima)* population of the Connecticut River, pp. 3–17. in Van Winkle, W. (Ed.), *Proceedings of the Conference on Assessing the Effects of Power-Plant-Induced Mortality on Fish Populations*. Pergamon Press, New York.
62. **Westernhagen, H. von, K. R. Sperling, D. Janssen, V. Dethlefsen, P. Cameron, R. Kocan, M. Landolt, G. Fürstenberg, and K. Kremling.** 1987a. Anthropogenic contaminants and reproduction in marine fish. *Ber. Biol. Anst. Helgol*. 3: 1–70; **Westernhagen, H. von, V. Dethlefsen, P. Cameron, and D. Janssen.** 1987b. Chlorinated hydrocarbon residues in gonads of marine fish and effects on reproduction. *Sarsia* 72: 419–422.
63. **Cross, J. N. and J. E. Hose.** 1988. Evidence for impaired reproduction in white croaker *(Genyonemus lineatus)* from contaminated areas off southern California. *Mar. Environ. Res*. 24: 185–188.
64. **Westernhagen, H. von.** 1988. Sub-lethal effects of pollutants on fish eggs and larvae, pp. 253–346. in Hoar, W. S. and D. J. Randall (Eds.), *Fish Physiology*. Academic Press, New York.
65. **Sindermann, C. J.** 1980. Pollution effects on fisheries — potential management activities. *Helgol. Meeresunters*. 33: 674–686.

8 Effects of Pollution on Shellfish Abundance

****

"BROWN TIDE" IN LONG ISLAND WATERS

The New York Air National Guard plane made its final crisis response flight for the month over the abnormally brown waters of Great South Bay on the outer coast of Long Island. It was late summer 1986 — the second year that bays on the Island had been discolored and choked by the massive growth of a planktonic microalgal species just recently identified as Aureococcus anorexefferens, an organism not previously known to cause blooms in that area of the coast. The findings from that day's survey were grim: current abundance of the toxic organism in the Bay was 1,000,000,000 algal cells per liter, similar to what it had been all summer, resulting in severe reduction of light penetration of the water. This had caused significant reduction in eel grass abundance and distribution in Long Island bays, and profound disturbances in other components of the shallow water ecosystem.

One of the animal species most affected by the algal bloom was the bay scallop, Argopecten irradians, the base for an important commercial shellfishery. In the summer of 1985, when the bloom began, most of the scallop larvae had died, resulting in a massive recruitment failure. New York scallop landings in that year were only 58% of the average for the preceding four years. Natural restocking was precluded by recurrence of the bloom in the summer of 1986, and its reappearance in some previously affected areas in 1987. The concurrent loss of critical eel grass habitat may serve to further inhibit reestablishment of the Long Island bay scallop fishery.

The problem was not confined to Long Island waters. The same alga, Aureococcus anorexefferens, bloomed from Narragansett Bay, Rhode Island, southward to Barnegat Bay, New Jersey, in 1985. It caused mortalities of mussels, Mytilus edulis, in excess of 95% in Narragansett Bay, and significant growth suppression in hard clams, Mercenaria

mercenaria, in Long Island bays. Mats of dead eel grass littered the shores in New York and New Jersey.

The problem did not disappear with the passage of time, either. A 1991 "NOAA News Bulletin" reported that an Aureococcus bloom had reoccurred in eastern Long Island Sound in June of that year, with cell densities eight times that known to harm marine animals. Bay scallop larvae were again assumed to have been killed by the early stages of the bloom.

A relationship of recurrent algal blooms such as these to modifications by humans of coastal/estuarine waters is often suggested, and some evidence exists. Nutrient enrichment from agricultural runoff, sewerage outfalls, and some industrial effluents may be involved, as may be the transport of toxic algae to new locations in ships' ballast water or with introduced marine animals. Whatever the causation, there seems to be a real increase in the frequency, intensity, and geographic area affected by algal blooms on a worldwide basis, and shellfish populations are among the impacted groups.

From "Field Notes of a Pollution Watcher"
(C. J. Sindermann, 1994)

****

INTRODUCTION

It may seem somewhat artificial to separate considerations of pollution effects on fish from those on shellfish, but reasons for the division exist:

1. Shellfish (except for some crustaceans) are nonmigratory or only weakly migratory, so the environment in which they are found as adults is the one to which they have been exposed for much of their existence. This also implies that the sedentary species are less likely to escape from habitats that have become toxic.
2. As a consequence of their sedentary habits, molluscan shellfish in polluted zones are prohibited from commercial exploitation. Closure of shellfish beds (roughly one-third of all productive acreage in the United States) is almost the equivalent of mortality unless the area can be cleaned up, or the existing stocks can be transferred to a clean site, or the polluted population can serve as a spawning source for adjacent stocks in marginal or acceptable areas.
3. Shellfish are easier to sample than fish, so mortalities are clearly detectable. This is especially true for molluscs, for which dead shells can be counted, and the interval following death can be determined by the extent of fouling of the inner shell.
4. Shellfish, as invertebrates, have a somewhat different array of physiological responses to pollutants than fish. Their immune system is fundamentally different and less specific than that of vertebrates, and the sensitivities of crustaceans to some contaminants such as DDT are significantly higher than those of fish.

5. Crustacean shellfish are subject to a unique spectrum of pollution-associated diseases such as shell erosion and black gills.
6. Bivalve molluscan shellfish are subject to highly lethal neoplastic diseases, possibly pollution associated, that in some species such as softshell clams may cause mortalities in excess of 90% of the local population.

Population effects on shellfish may be categorized as *direct* or *indirect*. Direct or acute effects result in mortalities when toxic contaminant levels in the environment exceed physiological limits (Figure 35). Indirect or sublethal effects can be more varied and interesting. Such subacute or chronic effects may include (but are not limited to):

- Effects on fecundity
  - Reduction in numbers of viable offspring
  - Behavioral changes reducing copulatory activity
  - Reduced fertilizing capacity of sperm
- Effects on physiological processes
  - Reduced growth
  - Sensory/neural pathology
  - Extended closure of valves
- Effects on body structures
  - Shell abnormalities in bivalves
  - Anomalies in appendages of crustaceans
- Effects on behavior
  - Escape reactions — as when clams leave their burrows in the presence of oil and are exposed to crab and bird predation

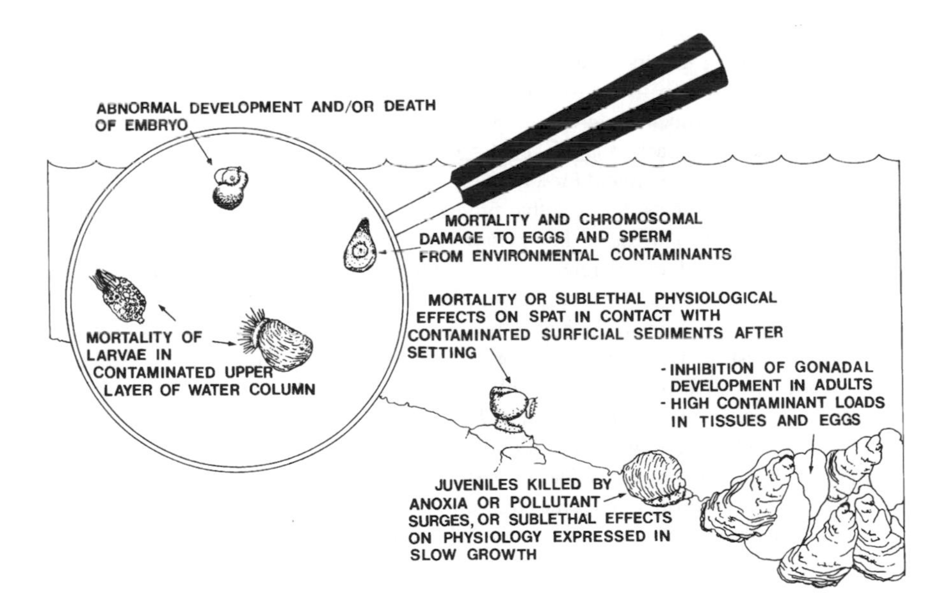

FIGURE 35. Contaminant effects on oyster life cycle stages.

To continue the process of subdivision, it now seems logical to separate the molluscs from the crustaceans in this treatment of pollution effects on shellfish, so that we do not constantly have to refer to exceptions. Some information is available on the quantitative effects of pollution on bivalve molluscs, but less on effects on crustacean resource species, and neither information set could be described as substantive.

EFFECTS OF POLLUTION ON MOLLUSCAN SHELLFISH POPULATIONS

Several examples of pollution-related events in molluscan shellfish should be indicative of the kinds of population data that are available — replete with percent mortalities, but deficient in estimates of population size and numbers of animals involved in the mortalities.

EFFECTS OF A LARGE-SCALE OIL SPILL

The tanker *Amoco Cadiz* was wrecked on the Brittany coast of France in 1978, spilling about 233,000 tons of crude oil. Effects on the commercially important oyster populations of the area have been examined repeatedly since that catastrophe, through a combination of field surveys and experimental introductions of clean oysters, *Ostrea edulis* and *Crassostrea gigas*.[1] Some of the findings relevant to quantitative effects of oil pollution are these:

1. Oyster mortalities of 20 to 50% were reported in the most heavily polluted sites during the first three months after the spill.
2. Lesions, principally necrosis or atrophy of the digestive tract epithelium, were observed frequently during 1978 and in several succeeding years, gradually decreasing to background levels by 1983.
3. No evidence was found for a relationship of oil pollution with increased prevalences or effects of two lethal protozoan pathogens — *Marteilia refringens* and *Bonamia ostreae* — in *O. edulis*.
4. Severe lesions of the reproductive system were seen in 1978 in the form of destruction of the duct epithelium, intense inflammatory reaction, and final atrophy. Total suppression of spawning of *O. edulis* was postulated for 1978.[2]
5. Neoplasms have been reported previously in *O. edulis,* but prevalences did not increase after the oil spill. Neoplasms with prevalences up to 40% were, however, observed in cockles, *Cerastoderma edule,* from the impacted area in 1983, but a relationship with oil pollution was not established.[3]
6. Hydrocarbon levels residual in oysters in 1985 — seven years after the spill — were still two to five times the values found in an unpolluted reference site, but pathological signs, seen earlier in the form of digestive gland and gonadal lesions, had disappeared. The so-called "indifference" of oysters to chronic

> petroleum pollution, as measured by the results of pathological examination, was suggested. This, however, is not consistent with findings from experimental exposures of another species of oyster, *Crassostrea virginica,* to chronic low levels of crude oil, in which severe alterations in the digestive tract and the gonads were seen.[4]

Continuing investigations of the effects of this single massive oil pollution event on the French coast on molluscan species have thus provided useful information about immediate as well as long-term damage.

Effects of Ocean Dumping on Offshore Molluscan Shellfish Stocks

Pollution effects on bivalve molluscs are particularly apparent in nearshore waters, but ocean dumping can contaminate offshore stocks as well. As a good example, the harvesting of shellfish was prohibited for many years in zones around the New York and Philadelphia ocean dump sites (Figure 36), because of demonstrated microbial and heavy metal contamination. Heavy metals well above background levels were found in surf clams from both dumpsite areas, and a 1979 report[5] pointed out other interesting events concerned with heavy metal pollution in the Middle Atlantic Bight:

1. Higher heavy metal levels (especially lead and cadmium) were found in ocean quahogs than in surf clams.
2. Highest levels of several metals in both species were found in the New York area, with decreasing values toward Cape Hatteras (a four- to five-fold decrease in silver, zinc, arsenic, copper, cadmium, and chromium). Mercury was below detection limits in most samples.
3. Some anomalies (higher values) were found in the vicinity of the Philadelphia sludge dump site.

To my knowledge, we have not yet seen any significant direct effect of ocean dumping on abundance of offshore molluscan resources. Effects thus far concern *lack of access to the resources in the impacted areas* (closed to harvesting); and a vague uneasiness about possible public health problems in peripheral areas and about sublethal effects of pollutant chemicals on the molluscs.

The problem of contamination of oceanic shellfish beds is not a simple one, as was pointed out in an impact analysis of the effects of ocean dumping at the so-called Philadelphia dump site.[6] Closure of an area by the Food and Drug Administration (FDA) of course results in removal of shellfish in that area from exploitation (an estimated 600,000 bushels of ocean quahogs in the closed area, for example), but there are subsidiary effects as well:

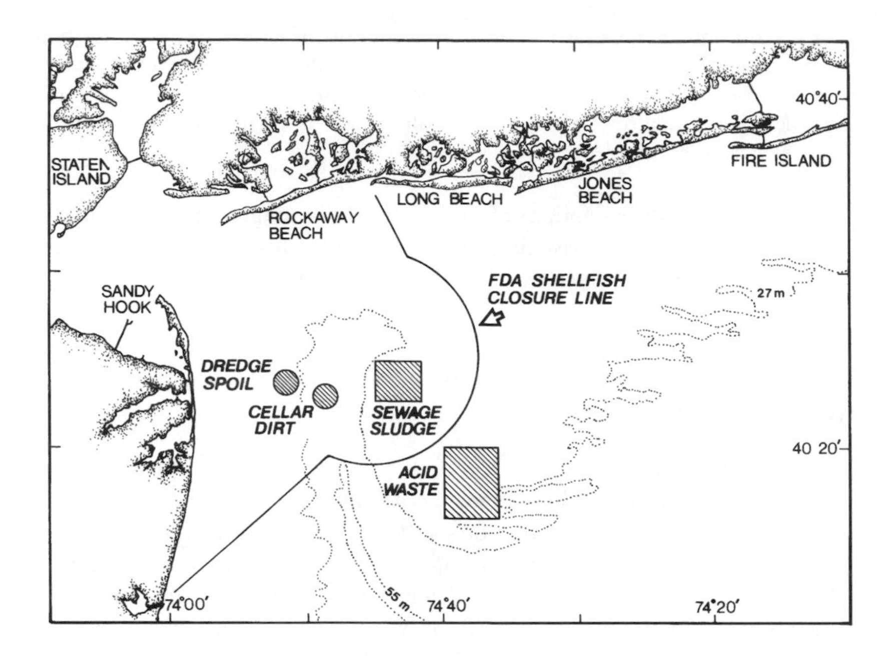

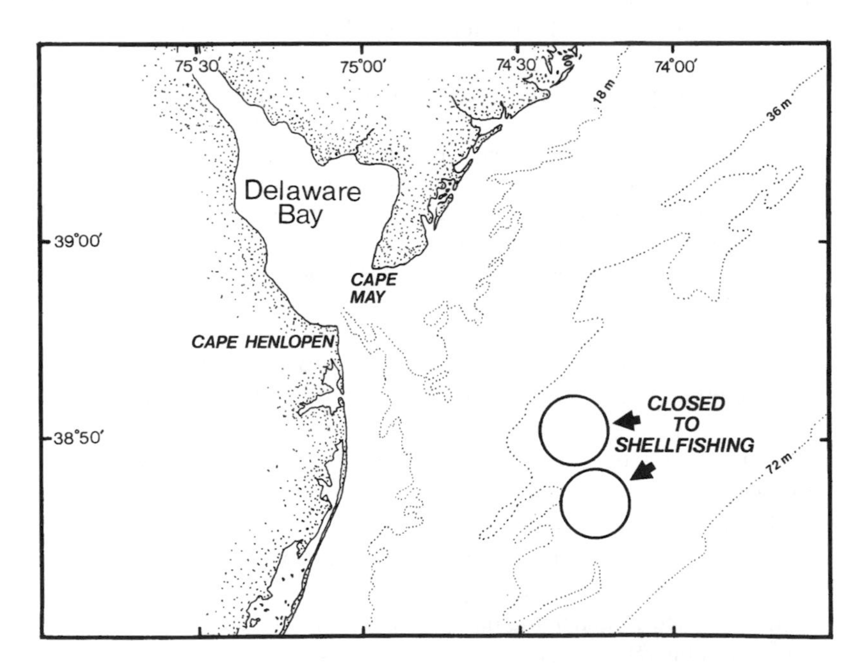

FIGURE 36. Dumpsites and dumpsite-affected areas of the New York Bight closed to shellfishing because of pollution. (*Top*) New York sites; (*bottom*) Philadelphia sites.

1. *Expansion of contaminated zones*
   Fecal coliform levels were high as far as 11 miles from the closed area off Maryland. Such findings raised the possibility of additional closures.
2. *Spread of the sludge plume*
   Prevailing current patterns and continued dumping can result in deposition of organic sediments far beyond actual dump site areas. The Philadelphia sludge dump site plume, for example, was depicted to extend well south of the closed area. However, no effects on abundance of shellfish were demonstrated in this plume area.

There is a large body of published experimental data on the acute effects of chemical contaminants (heavy metals, halogenated hydrocarbons, and petroleum components) on life history stages of marine animals. Localized effects of acute contamination by oil spills on coastal/estuarine shellfish beds have been documented, but evidence does not exist for comparable acute effects of pollutants on the abundance of offshore species. Long-term effects of contaminants on survival of life stages of shellfish have also been demonstrated experimentally, but application of findings to natural populations must be made with great care and conservatism, particularly when dealing with deeper-water species.

Thus, there are several direct and potential impacts of ocean dumping on oceanic shellfish harvesting. Economic impacts of closures can be estimated, but *biological* impacts on abundance of commercial shellfish have yet to be determined and will be very difficult to determine until we have better understanding of the natural factors that affect population abundance.

Pollution-Induced Mortalities in Near-Shore Molluscan Shellfish Stocks

The combined impacts of reduction in growth rates and increase in mortality rates of molluscs are common phenomena in polluted habitats. Several studies have demonstrated these effects. As an example, the growth rates of soft clams, *Mya arenaria,* declined by 65% when clams were transplanted from a clean site to an oil-polluted site on the Maine coast.[7] Survival of the transplants at the polluted site was only 12%, compared with 78% at a reference site.

In another study, slow growth and high mortality in black abalone, *Haliotis cracherodii,* from a polluted area of the California coast were reported.[8] Populations from the Los Angeles County sewer outfall site were compared with those from Santa Catalina Island. The polluted outfall site yielded samples of abalone that were starving, had eroded shells, and died at a much higher rate than did their island relatives. Individuals transplanted from the clean site to the outfall area failed to grow and eventually died from undetermined causes.

Findings were reminiscent of the results of earlier studies of fish of several species sampled at outfall sites in California waters, in which lesions, emaciation, oral tumors, and exophthalmia were noted.

Then, beginning in the mid-1980s, black abalone landings in southern California, already depressed as a consequence of overfishing, began to decline still further as mass mortalities from unknown causes occurred.[9] Within five years, the black abalone virtually disappeared from most of its range south of Point Conception. To the present time (1995), the cause(s) of the population crash is still undetermined, although pollution, disease, and high water temperatures have all been suspected.[10]

Molluscan Neoplasia and Associated Mortalities

The research literature on diseases of bivalve molluscs has been dominated since 1969 by reports of neoplasms ("cancers"), sometimes at epizootic levels, in clams, oysters, mussels, and cockles (as summarized briefly in Chapter 4). The neoplastic condition is progressive, invasive, and lethal. It has been associated with mortalities of oysters, mussels, and clams, and — at least in the case of softshell clams on the east coast of the U.S. — is transmissible and possibly of infectious etiology, based on field observations and experimental studies.[11] Because the disease occurs in a number of economically important bivalves and because epizootic levels may affect population abundance, it seems relevant to reexamine the condition in some detail, as it occurs in each major bivalve group, and especially as it relates to abundance of populations. The cause is unknown, except in the case of softshell clams, *Mya arenaria,* in which a retrovirus has been implicated.[12] In other studies, oil pollution has been suggested as a possible contributing factor,[13] but the evidence must be described as conflicting and inconclusive. One group of investigators[14] found a correlation between tissue levels of polycyclic aromatic hydrocarbons (PAH) and neoplasia in mussels from the Oregon coast, but French studies at the site of the massive *Amoco Cadiz* oil spill of 1978 disclosed no relationship between oil pollution and neoplasms (sarcomas) in European flat oysters.[15] Polychlorinated biphenyl (PCB) pollution of growing areas has been implicated in enhanced prevalences of neoplasia in softshell clams.[16] The weight of evidence thus far suggests that any direct association of molluscan neoplasms with specific pollutants is premature, but that environmental stressors (possibly in concert with an infectious agent) may contribute to the development of epizootic levels.

The greatest amount of literature on neoplasms of bivalve molluscs is concerned with studies of softshell clams. The emphasis has been a result of (1) interest in the neoplasms as possible biological indicators of oil and PCB pollution and (2) the discovery of epizootic levels of neoplasia in some geographic populations of the species. The story to date is a fascinating one.

An oil spill at Searsport, Maine, in 1971 provided the impetus for the initial histopathological studies that disclosed the presence of neoplasia in clams. Examination of samples from the spill site revealed high prevalences of gonadal tumors (4.3%), characterized by numerous mitotic figures, multinucleated giant cells, lobed irregular nuclei, and invasion of the connective tissue around gonadal follicles.[17] A subsequent study of softshell clams from other coastal areas — particularly from another oil-polluted site, at Brunswick, Maine — disclosed the presence of a second type of invasive neoplasm, thought to originate in blood cell–forming (hematopoietic) tissues, and characterized by large lobed nuclei, numerous and irregular mitotic figures, and extensive invasion of connective tissue throughout the body.[18] A still later survey of 10 New England softshell clam populations resulted in the findings of neoplasms of this second type with prevalences of 0 to 64%.[19] Advanced neoplastic disease was associated, in laboratory holding studies, with mortalities.

New information on the course of the disease was provided in 1982.[20] Observations of laboratory-held neoplastic and normal clams indicated that:

- Significantly higher mortalities occurred in neoplastic clams, with survival rates inversely related to the severity of the disease.
- The disease followed one of three courses: (1) it could be progressive, leading to increased tissue invasion and death (50% of infected individuals); (2) it could become chronic (40% of infected individuals); or (3) it could diminish in severity or disappear (10% of infected individuals).

Transmissibility of the neoplasms in softshell clams was demonstrated experimentally, and the investigators also suggested that environmental stress may have increased susceptibility to neoplasia.[21] Subsequent to this study, a B-type retrovirus was reported to have been isolated from neoplastic clams, and the disease was reported to have been reproduced by inoculating purified virus preparations.[22] The virus was not, however, identified definitively, nor was it fully characterized.

Most of the earlier research funding for studies of softshell clam neoplasia resulted from interest in the possible association of the disease with oil pollution in northern New England waters. Beginning in the early 1980s, however, interest shifted to new epizootics of neoplasms in clams from New Bedford Harbor, Massachusetts, and Chesapeake Bay. In New Bedford Harbor, an area noted for high levels of PCBs and other industrial pollutants, prevalences of neoplasms were reported to approach 90% in some samples.[23] Detectable levels of PCBs were found in neoplastic cells from New Bedford clams, but not in blood cells of normal clams from the same sites, a possible reflection of altered fat metabolism in neoplastic cells.

A two and a half year survey (1983 to 1985) of the occurrence of neoplasia in softshell clams from Long Island Sound, New York, disclosed a low overall prevalence of slightly over 1%, but with pronounced annual and seasonal variations.[24] Peak prevalences (as high as 60%) occurred in late fall and winter in some of the sites sampled.

Concern about epizootic levels of neoplasia in softshell clams from Chesapeake Bay during the past decade (1984 to 1994) stems from a more immediate problem: the potential for major negative effects of disease-induced mortalities on commercial clam abundance. The neoplastic disease of clams was not seen in the bay prior to 1979 and was a rare occurrence until 1983, when it first appeared at epizootic levels. Prevalences remained high until April 1985. Seasonal progression of disease intensity was observed; light infections in December progressed to advanced and terminal stages by the following April.[25]

Chronic disease and remissions, as seen in New England clam populations,[26] were not evident in Chesapeake Bay investigations. Laboratory observations indicated 100% mortality of diseased Chesapeake Bay clams from April to June, and field prevalences dropped to 0% in June, presumably because of the deaths of animals with advanced disease. The Chesapeake Bay investigators[27] concluded that high prevalences and advancing stages of the disease seen in natural populations could be precursors of significant mortalities in commercial stocks. Those researchers also suggested an infectious etiology and pointed out the possibility that the infectious agent may have been introduced into the bay with clams transferred from New England prior to 1979.

The recent (1983 to 1985) outbreak of neoplastic disease in softshell clams from Chesapeake Bay is of particular interest for several reasons:

1. Surveys that began in 1969 demonstrated that the disease condition did not occur in the population until 1979, but that prevalences increased to 40% in some 1983 samples and to 91% by 1985. An introduced infectious agent was strongly indicated by the data.
2. Progression of the disease through well-defined stages was followed simultaneously in field and experimental populations. Mortalities were correlated with advanced stages of the disease.
3. Transplantability of the disease was demonstrated experimentally by the injection of neoplastic cells; 95% of injected clams developed neoplasms; and advanced infections were lethal in 1 to 4 months.
4. *Data from the clam fishery indicated a severe impact of neoplastic disease on stocks*. As an example, a one-year survey of one clam population (Swan Point, October 1984 to October 1985) disclosed an initial high prevalence of the disease, which was followed by high mortalities and a concomitant decline in prevalence. The density of clams and the catch-per-unit effort decreased, and the site became commercially unproductive in the following year (1986). It seems relevant that the production of clams in Maryland waters of the Chesapeake Bay declined by 50% from 1983 to 1985, the period encompassed by the epizootic.[28]

Neoplasia has also been reported in several species of oysters of the genera *Ostrea* and *Crassostrea*. Neoplastic disease, with characteristics like those seen in clams and other bivalves, has been suggested as the cause of recurring mortalities of European flat oysters, *Ostrea edulis,* in Spain and in Yugoslavia.[29] Mortalities varied yearly and geographically, but were in the 20 to 90% range in Yugoslavia and 60 to 80% in Spain. Oysters sampled from affected areas showed high levels of cellular infiltration of connective tissues. Neoplasms have also been reported in native (Olympia) oysters, *Ostrea lurida,* from the Pacific Northwest.[30] In the most extensive study, prevalences approximated only 2% and no mortalities were attributed to the disorder.

Eastern oysters, *Crassostrea virginica,* from Chesapeake Bay — 12 of 20,000 specimens examined over a 14-year period — were diagnosed as having neoplasms.[31] One study of this species included examination of over 100 small laboratory-reared inbred subpopulations, two of which exhibited high prevalences of neoplasms. One of these laboratory populations suffered high mortalities that seemed to be directly related to high prevalences of neoplasms in the dying oysters. The study suggested possible genetic differences in susceptibility to neoplasia among the inbred oyster populations.

A neoplastic disease of blue mussels, *Mytilus edulis,* from Yaquina Bay, Oregon, was first described in 1969.[32] Characteristics of the disease were similar to those of the neoplasms of oysters and clams: unusually large cells with nuclei two to four times larger than normal blood cells, abundant mitotic figures, and the presence of aggregations of abnormal cells in the connective tissue. Mussels with advanced disease were emaciated, with degenerative changes suggesting lethality. A subsequent five-year survey of the disorder (1976 to 1981) in Yaquina Bay disclosed a seasonal pattern of prevalences (with a peak from January to March), marked geographic variation in disease abundance and a mean prevalence of 9.8% in the subpopulation with the highest levels of the disease.[33] In the words of the investigator, though, "There was no strong indication that significant mortalities occurred as a result of the condition." The cause of the disease condition was undetermined, although a concurrent chemical study disclosed a statistical association between tissue levels of polycyclic aromatic hydrocarbons (PAHs) and the disorder.[34] Counterpart surveys of the disease in British Columbia (Canada) waters disclosed an overall prevalence of 12.7% in mussels, with epizootic levels (12 to 29%) in some locations.[35]

A recent experimental study of the disorder in mussels from an aquaculture population being reared in Puget Sound, Washington disclosed that the condition was usually progressive and fatal, but that 20% of the experimental group showed signs of remission.[36] Transplantation of neoplasms to unaffected mussels was achieved, using inoculations of intact neoplastic cells. Transmission by inoculation with cell-free homogenates and by aquarium exposure of disease-free individuals to neoplastic animals was also reported.[37]

To summarize this abundant information, prevalences of neoplastic diseases indicative of epizootics have been reported in a number of bivalve species and in widely separated locations. A few outbreaks have been well documented:

- Neoplasms have been observed at epizootic levels in blue mussels, *Mytilus edulis,* from Yaquina Bay, Oregon,[38] and from the British coast.[39]
- Neoplasms at epizootic levels have been reported in local populations of duck clams, *Macoma balthica,* from Chesapeake Bay.[40]
- Two different types of neoplasms have been seen in softshell clams, *Mya arenaria,* from several locations on the New England coast in the 1970s,[41] and one type (often referred to as hematopoietic neoplasia) has been subsequently reported at epizootic levels in clams from New Bedford Harbor, Massachusetts, and Chesapeake Bay.[42]

The possible association of epizootic levels of neoplasms with pollution must be described at present as "tenuous," although some of the literature suggests possible relationships, especially with petroleum and some of its derivatives.

EFFECTS OF POLLUTION ON CRUSTACEAN SHELLFISH POPULATIONS

****

THE GREAT IXTOC/GULF OF MEXICO OIL SPILL

This had to be a petroleum-related environmental event of catastrophic proportions, even for an area such as the Gulf of Mexico, where contamination from offshore oil production has been a common occurrence. Ixtoc 1, a Mexican exploratory well located in the Bay of Campeche off the Yucatan Peninsula, had blown out on June 3, 1979, and was initially gushing an estimated 1 million gallons of oil per day into the Gulf. The oil slick had reached the beautiful beaches of Padre Island off the Texas coast by late August of that year.

Totally by coincidence, a major meeting of aquatic scientists and entrepreneurs — the World Aquaculture Society — was being held in late August, in Corpus Christi, not far away from those by then-despoiled beaches. Ad hoc field trips to the impacted areas were quickly organized, to allow meeting participants to confront the reality of what is a perennial concern of aquaculturists: spilled oil polluting productive waters. The scene on the beaches was overwhelming: a black rim of oil stretching along the shore to the horizons; overpowering oil fumes and oily spray; great globs of brown goo called "mousse" sloshing in the surf and coating the sand; front-end loaders charging up and down the beach in a futile effort to scoop up the oil-soaked sand; NOAA crisis response teams in helicopters and boats, assessing the condition and movement of the oil — all combining to create an unreal science fiction-like stage set. At risk, in many people's minds, was the major Gulf shrimp fishery. Predictions of

disaster for the shrimp industry from mass mortalities and tainted catches were universal, and the future well-being of this estuarine-dependent species in a contaminated environment was in question.

The oil flow was reduced by late summer, and the well was finally capped in March 1980, after releasing an estimated 130 million gallons of oil into the Gulf (some estimates are much higher). The aquaculturists flew home from their meeting in Corpus Christi; a few reports of tainted shrimp catches were highly publicized; the oil slick dissipated; and the event faded into history almost without a trace. Possible impacts of the spill on subsequent year classes of shrimp and other resource species of the Gulf were examined by fisheries scientists, with equivocal results, because of the difficulty in separating effects of the oil from those of other environmental factors that influence population abundance. Annual Gulf shrimp landings since 1979 have fluctuated around 100,000 tons. As time passed, it even became difficult to locate bottom deposits of oil, except in the immediate vicinity of the well head.

So from the perspective of the human intruder, the event seemed catastrophic, but from that of the resource species and the entire Gulf ecosystem it probably represented a relatively mild perturbation, possibly of lesser overall significance than the passage of an average-strength hurricane.

From "Field Notes of a Pollution Watcher"
(C. J. Sindermann, 1990)

****

Moving on to examples of pollution events and their consequences in marine crustacean populations, it is possible to find numerous references to pollution-associated abnormalities and their possible impacts on survival, but, as with the molluscs, little specific data on numbers of animals killed. With few exceptions, estimates of percent mortality are not accompanied by estimates of population size.

Crustaceans have their own suite of pollution-or stress-associated diseases. Prominent among them is a condition known as "shell disease" caused by destruction by bacterial action of portions of the chitinous exoskeleton; a related abnormality is described as "black gill disease."

Shell Disease in Crustaceans

Shell disease has a reasonably good association with stressful conditions and is frequently associated with badly degraded estuarine and coastal waters. Also called "exoskeletal disease," "shell erosion," or "black spot," it can be considered in some ways as the invertebrate analog of fin erosion in fish (described in Chapter 3).

Lobsters, *Homarus americanus,* and rock crabs, *Cancer irorratus,* from grossly polluted areas of the New York Bight were found in an extensive study in 1975 to be abnormal,[43] with appendage and gill erosion a most common sign. Skeletal erosion occurred principally on the tips of the walking legs, the ventral sides of chelipeds, exoskeletal spines, gill lamellae, and around areas of exoskeletal articulation, where contaminated sediments could accumulate. Gills of crabs and lobsters sampled at the dump sites were usually clogged with detritus, possessed a dark brown coating, contained localized thickenings, and displayed areas of erosion and necrosis. Similar disease signs were produced experimentally in animals held for six weeks in aquaria containing sediments from sewage sludge or dredge spoil disposal sites. Initial discrete areas of erosion became confluent, covering large areas of the exoskeleton, and often parts of appendages were lost. The chitinous covering of the gill filaments was also eroded, and often the underlying tissues became necrotic.

Dead and moribund crabs and lobsters have been reported on several occasions by divers in the New York Bight apex, and dissolved oxygen concentrations near the bottom during the summer often approached zero.[44] Low oxygen stress, when combined with gill fouling, erosion, and necrosis, could readily lead to mortality.[45]

Mortalities due to the shell disease syndrome have been reported for several crustacean species. In lobsters, death may be the consequence of progressive erosion and destruction of gill membranes with resultant reduced oxygen uptake, especially in hypoxic situations (near sewage sludge dump sites, for example). In shrimp, failure to complete ecdysis because of shell adhesions has been identified as a causal factor. In these and other species, death may also result from secondary infections, after the exoskeletal barrier has been breached, especially in the presence of high populations of facultative pathogens. Additionally, it is quite likely that severely affected animals would be more vulnerable to predators (Figure 37).

Black Gill Disease of Crustaceans

The shell disease syndrome in crustaceans from polluted habitats is intimately associated with another general disease sign or syndrome — the so-called "black gill syndrome," characterized by sediment accumulation between gill lamellae, accompanied by darkening of filaments. Substrate is provided for microbial growth, and the syndrome is further characterized by chitin deterioration and gill tissue necrosis.[46]

A recent study of shell disease and black gill disease in lobsters from Massachusetts waters showed similar trends for both diseases, with the highest prevalences (up to 50%) in samples from the most polluted sites, particularly Boston Harbor and Buzzards Bay.[47] Prevalences in more exposed, deeper water

FIGURE 37. Severe shell disease in a blue crab (*Callinectes sapidus*). (Photograph courtesy of Dr. D. W. Engel.)

sites were generally low (Cape Ann, Cape Cod Bay, outer Cape Cod, and Eastern Shore). Mortalities were not observed, but population impacts were postulated, based on increased vulnerability to hypoxia of lobsters with fouled or necrotic gills.

Chronic Habitat Contamination

In addition to pollution-related diseases and possible associated mortalities, crustacean populations may be reduced by acute and chronic contamination of habitats. As an example, a long-term study of the effects of a 1969 fuel oil spill in Buzzards Bay, Massachusetts, on populations of fiddler crabs, *Uca pugnax,* disclosed direct mortalities of adults where surface sediment oil concentrations exceeded 1000 ppm.[48] Additionally, oiled sediments (in excess of 200 ppm) were toxic to overwintering juvenile crabs. Recruitment was below normal for several years after the spill and population densities were depressed for seven years before recovery began. Behavioral disorders caused by contamination, particularly locomotor impairment, abnormal burrow construction, and abnormal mating activities, were considered by the investigators to have contributed to mortalities.

CONCLUSIONS

These few examples of the possible impacts of pollution on shellfish populations lead to the correct conclusion that quantitative information about effects on population abundance is scarce and inadequate — even worse than that for fish. After examining enough published reports, it seems that the best data concerns short-term acute pollution events such as oil spills, but that data from comprehensive long-term studies of subacute chronic effects on abundance are more difficult to locate.

In an ideal world of unlimited research funding and ready availability of competent investigators, determination of the effects of pollution on shellfish abundance should include several critical components:

- An estimate of the size and distribution of the population before the event, and its trends in abundance to the present time
- An estimate of the total area of pollution effects, in degrees of severity, from mortalities to no effect
- Data from experimental studies of the effects of exposure to various levels of the principal contaminants on different life cycle stages and for varying periods of time
- A simultaneous examination of all other potential environmental influences on shellfish survival and abundance
- A simultaneous study of shellfish survival and abundance in a relatively unpolluted reference (control) area
- A concurrent simulation modeling effort closely allied with and interactive with field and laboratory studies
- A guarantee of continuity of the project for at least one decade, and preferably longer

This listing is of course the stuff of which dreams are made. Meanwhile, we must satisfy ourselves with short-term and otherwise incomplete data sets, with mostly poorly documented observations of mortalities in localized acute pollution events, and with tantalizing speculations about the long-term chronic effects of coastal pollution on shellfish populations.

REFERENCES

1. **Berthou, F., G. Balouet, G. Bodennec, and M. Marchand.** 1987. The occurrence of hydrocarbons and histopathological abnormalities in oysters for seven years following the wreck of the *Amoco Cadiz* in Brittany (France). *Mar. Environ. Res.* 23: 103–133.
2. **Neff, J. M. and W. E. Haensly.** 1982. Long-term impact of the *Amoco Cadiz* oil spill on oysters *Crassostrea gigas* and plaice *Pleuronectes platessa* from Aber-Benoit and Aber-Wrach, Brittany, France, pp. 269–328. in Ecological Study of the *Amoco Cadiz* Oil Spill. NOAA-CNEXO Rep., U.S. Department of Commerce, Washington, D.C.
3. **Poder, M. and M. Auffret.** 1986. Sarcomatous lesions in the cockle *Cerastoderma edule*. *Aquaculture* 56: 1–8.
4. **Barszcz, C. A., P. P. Yevich, L. R. Brown, J. D. Yarbrough, and C. D. Minchew.** 1978. Chronic effects of three crude oils on oysters suspended in estuarine ponds. *J. Environ. Pathol. Toxicol.* 1: 879–895.

5. **Wenzloff, D. R., R. A. Greig, A. S. Merrill, and J. W. Ropes.** 1979. A survey of heavy metals in two bivalve molluscs of the mid-Atlantic coast of the United States. *Fish. Bull.* 77: 280–285.
6. **Forste, R. H. and R. G. Rinaldo.** 1977. Estimated impacts of the Philadelphia dump site on the sea clam fishery. Md. Dept. Natur. Resour. Doc. FA-MRR-77–1, 22 pp.
7. **Dow, R. L.** 1975. Reduced growth and survival of clams transplanted to an oil spill site. *Mar. Pollut. Bull.* 6: 124–125.
8. **Young, P. H.** 1964. Some effects of sewer effluent on marine life. *Calif. Fish Game* 50: 33–41.
9. **Davis, G. E.** 1993. Mysterious demise of southern California black abalone, *Haliotis cracherodii* Leach, 1814. *J. Shellfish Res.* 12: 183–184.
10. **Vanblaricom, G. R., J. L. Ruediger, C. S. Friedman, D. D. Woodard, and R. P. Hedrick.** 1993. Discovery of withering syndrome among black abalone *Haliotis cracherodii* Leach 1814, populations at San Nicolas Island, California. *J. Shellfish Res.* 12: 185–188; **Miller, A. C. and S. E. Lawrenz-Miller.** 1993. Long-term trends in black abalone, *Haliotis cracherodii* Leach, 1814, populations along the Palos Verdes Peninsula, California. *J. Shellfish Res.* 12: 195–200.
11. **Brown, R. S., R. E. Wolke, S. B. Saila, and C. W. Brown.** 1977. Prevalence of neoplasia in 10 New England populations of the soft-shell clam *(Mya arenaria)*. *Ann. N.Y. Acad. Sci.* 298: 522–534; **Farley, C. A., S. V. Otto, and C. L. Reinisch.** 1986. New occurrence of epizootic sarcoma in Chesapeake Bay soft shell clams *(Mya arenaria)*. *Fish. Bull.* 84: 851–857; **Elston, R. A., S. K. Allen, E. W. Radnay, and M. L. Kent.** 1987. Experimental studies on hemic proliferative disease in the mussel, *Mytilus edulis*. *Prog. Abstr. 20th Annu. Meet. Soc. Invertebr. Pathol.*, p. 85.
12. **Cooper, K. R. and P. W. Chang.** 1982. A review of the evidence supporting a viral agent causing an haematopoietic neoplasm in the soft shell clam *Mya arenaria*. *Proc. 15th Int. Colloq. Invertebr. Pathol.*, pp. 271–272.
13. **Farley, C. A.** 1969. Probable neoplastic disease of the hematopoietic system in oysters, *Crassostrea virginica* and *Crassostrea gigas*. *Natl. Cancer Inst. Monogr.* 31:541–555; **Yevich, P. P. and C. A. Barszcz.** 1977. Neoplasia in soft-shell clams *(Mya arenaria)* collected from oil-impacted sites. *Ann. N.Y. Acad. Sci.* 298: 409–426; **Mix, M. C., R. L. Schaffer, and S. J. Hemingway.** 1981. Polynuclear aromatic hydrocarbons in bay mussels *(Mytilus edulis)* from Oregon, pp. 167–177. in Dawe, C.J., J.C. Harshbarger, S. Kondo, T. Sugimura, and S. Takayama (Eds.), *Phyletic Approaches to Cancer.* Japan Science Society Press, Tokyo.
14. **Mix, M. C., S. J. Hemingway, and R. L. Schaffer.** 1982. Benzo(a)pyrene concentrations in somatic and gonad tissues of bay mussels, *Mytilus edulis*. *Bull. Environ. Contam. Toxicol.* 28:46–51.
15. **Balouet, G., M. Poder, A. Cahour, and M. Auffret.** 1986. Proliferative hemocytic condition in European flat oyster *(Ostrea edulis)* from Breton coasts: a six-year survey. *J. Invertebr. Pathol.* 48: 208–215; **Poder, M. and Auffret, M.** 1986. Sarcomatous lesion in the cockle *Cerastoderma edule*. I. Morphology and population survey in Brittany, France. *Aquaculture* 58: 1–8.

16. **Reinisch, C. L., A. M. Charles, and A. M. Stone.** 1984. Epizootic neoplasia in soft shell clams collected from New Bedford Harbor. *Hazard. Waste* 1: 73–81.
17. **Barry, M. M. and P. P. Yevich.** 1975. The ecological, chemical and histopathological evaluation of an oil spill site. III. Histopathological studies. *Mar. Pollut. Bull.* 6: 171–173.
18. **Yevich, P. P. and C. A. Barszcz.** 1976. Gonadal and hematopoietic neoplasms in *Mya arenaria. Mar. Fish. Rev.* 38(10): 42–43; **Yevich, P. P. and C. A. Barszcz.** 1977. Neoplasia in soft-shell clams *(Mya arenaria)* collected from oil-impacted sites. *Ann. N.Y. Acad. Sci.* 298: 409–426.
19. **Brown, R. S., R. E. Wolke, and S. B. Saila.** 1976. A preliminary report on neoplasia in feral populations of the soft-shell clam *Mya arenaria:* prevalence, histopathology and diagnosis. *Proc. 9th Int. Colloq. Invertebr. Pathol.*, pp. 151–158; **Brown, R. S., R. E. Wolke, S. B. Saila, and C. W. Brown.** 1977. Prevalence of neoplasia in 10 New England populations of the soft-shell clam *(Mya arenaria). Ann. N.Y. Acad. Sci.* 298: 522–534.
20. **Cooper, K. R., R. S. Brown, and P. W. Chang.** 1982a. Accuracy of blood cytological screening techniques for the diagnosis of a possible hematopoietic neoplasm in the bivalve mollusk, *Mya arenaria. J. Invertebr. Pathol.* 39: 281–289; **Cooper, K. R., R. S. Brown, and P. W. Chang.** 1982b. The course and mortality of a hematopoietic neoplasm in the soft-shell clam, *Mya arenaria. J. Invertebr. Pathol.* 39: 149–157.
21. **Brown, R. S.** 1980. The value of the multidisciplinary approach to research on marine pollution effects as evidenced in a three-year study to determine the etiology and pathogenesis of neoplasia in the soft-shell clam, *Mya arenaria. Rapp. P.-V. Reun., Cons. Int. Explor. Mer* 179: 125–128.
22. **Appeldoorn, R. S. and J. J. Oprandy.** 1980. Tumors in soft-shell clams and the role played by a virus. *Maritimes* 24(1): 4–6; **Oprandy, J. J., P. W. Chang, A. D. Pronovost, K. R. Cooper, R. S. Brown, and V. J. Yates.** 1981. Isolation of a viral agent causing hematopoietic neoplasia in the soft-shell clam, *Mya arenaria. J. Invertebr. Pathol.* 38: 45–51.
23. **Reinisch, C. L., A. M. Charles, and A. M. Stone.** 1984. Epizootic neoplasia in soft shell clams collected from New Bedford Harbor. *Hazard. Waste* 1: 73–81.
24. **Brousseau, D. J.** 1987. Seasonal aspects of sarcomatous neoplasia in *Mya arenaria* (soft-shell clam) from Long Island Sound. *J. Invertebr. Pathol.* 50: 269–276.
25. **Farley, C. A., S. V. Otto, and C. L. Reinisch.** 1986. New occurrence of epizootic sarcoma in Chesapeake Bay soft shell clams *(Mya arenaria). Fish. Bull.* 84: 851–857.
26. **Cooper, K. R., R. S. Brown, and P. W. Chang.** 1982b. The course and mortality of a hematopoietic neoplasm in the soft-shell clam, *Mya arenaria. J. Invertebr. Pathol.* 39: 149–157.
27. **Farley, C. A., S. V. Otto, and C. L. Reinisch.** 1986. New occurrence of epizootic sarcoma in Chesapeake Bay soft shell clams *(Mya arenaria). Fish. Bull.* 84: 851–857.
28. **Farley, C. A.** 1989. Selected aspects of neoplastic progression in molluscs, pp. 24–31. in Kaiser, H.E. (Ed.), *Progressive Stages of Malignant Growth/Development.* Vol. I, Part 5: *Comparative Aspects of Tumor Progres-*

sion. Nijhoff, New York; **Farley, C. A., D. L. Plutschak, and R. F. Scott.** 1991. Epizootiology and distribution of transmissible sarcoma in Maryland soft-shell clams, *Mya arenaria,* 1984–1988. *Environ. Health Perspect.* 90: 35–41.

29. **Alderman, D. J., P. van Banning, and A. Perez-Colomer.** 1977. Two European oyster *(Ostrea edulis)* mortalities associated with an abnormal hemocytic condition. *Aquaculture* 10: 335–340.
30. **Jones, E. and A. K. Sparks.** 1969. An unusual histopathological condition in *Ostrea lurida* from Yaquina Bay, Oregon. *Proc. Natl. Shellfish. Assoc.* 59: 11; **Farley, C. A. and A. K. Sparks.** 1970. Proliferative diseases of hemocytes, endothelial cells, and connective tissue cells in mollusks. *Bibl. Haematol.* 36: 610–617; **Mix, M. C., H. J. Pribble, R. T. Riley, and S. P. Tomasovic.** 1977. Neoplastic disease in bivalve mollusks from Oregon estuaries with emphasis on research on proliferative disorders in Yaquina Bay oysters. *Ann. N.Y. Acad. Sci.* 298: 356–373.
31. **Couch, J. A.** 1969. An unusual lesion in the mantle of the American oyster, *Crassostrea virginica*. *Natl. Cancer Inst. Monogr.* 31: 557–562; **Farley, C. A.** 1969. Probable neoplastic disease of the hematopoietic system in oysters, *Crassostrea virginica* and *Crassostrea gigas*. *Natl. Cancer Inst. Monogr.* 31: 541–555; **Frierman, E. M. and J. D. Andrews.** 1976. Occurrence of hematopoietic neoplasms in Virginia oysters. *J. Natl. Cancer Inst.* 56: 319–324; **Harshbarger, J. C., S. V. Otto, and S. C. Chang.** 1979. Proliferative disorders in *Crassostrea virginica* and *Mya arenaria* from the Chesapeake Bay and intranuclear virus-like inclusions in *Mya arenaria* with germinomas from a Maine oil spill site. *Haliotis* 8: 243–248.
32. **Farley, C. A.** 1969. Sarcomatoid proliferative disease in a wild population of blue mussels *(Mytilus edulis)*. *J. Natl. Cancer Inst.* 43: 509–516.
33. **Mix, M. C.** 1983. Haemic neoplasms of bay mussels, *Mytilus edulis* L. from Oregon: occurrence, prevalence, seasonality and histopathological progression. *J. Fish Dis.* 6: 239–248.
34. **Mix, M. C., S. R. Trenholm, K. I. King.** 1979. Benzo*(a)*pyrene body burdens and the prevalence of cellular proliferative disorders in mussels, *Mytilus edulis,* from Yaquina Bay, Oregon. in *Pathobiology of Environmental Pollutants — Animal Models and Wildlife as Monitors*, National Academy of Science, Washington, D.C.; **Mix, M. C., R. L. Schaffer, and S. J. Hemingway.** 1981. Polynuclear aromatic hydrocarbons in bay mussels *(Mytilus edulis)* from Oregon, pp. 167–177, in Dawe, D.J., J.C. Harshbarger, S. Kondo, T. Sugimura, and S. Takayama (Eds.), *Phyletic Approaches to Cancer.* Japanese Science Society Press, Tokyo.
35. **Cosson-Mannevy, M. A., C. S. Wong, and W. J. Cretney.** 1984. Putative neoplastic disorders in mussels *(Mytilus edulis)* from southern Vancouver Island waters, British Columbia. *J. Invertebr. Pathol.* 44: 151–160.
36. **Elston, R. A., M. L. Kent, and A. S. Drum.** 1988. Progression, lethality, and remission of hemic neoplasia in the bay mussel, *Mytilus edulis*. *Dis. Aquat. Org.* 4: 135–142.
37. **Elston, R. A., M. L. Kent, and A. S. Drum.** 1988. Transmission of hemic neoplasia in the bay mussel, *Mytilus edulis,* using whole cells and cell homogenate. *Dev. Comp. Immunol.* 12: 719–729.

38. **Farley, C. A. and A. K. Sparks.** 1970. Proliferative diseases of hemocytes, endothelial cells, and connective tissue cells in mollusks. *Bibl. Haematol.* 36: 610–617; **Cosson-Mannevy, M. A., C. S. Wong, and W. J. Cretney.** 1984. Putative neoplastic disorders in mussels *(Mytilus edulis)* from southern Vancouver Island waters, British Columbia. *J. Invertebr. Pathol.* 44: 151–160.
39. **Green, M. and D. J. Alderman.** 1983. Neoplasia in *Mytilus edulis* from United Kingdom waters. *Aquaculture* 30: 1–10.
40. **Christensen, D. J., C. A. Farley, and F. G. Kern.** 1974. Epizootic neoplasms in the clam *Macoma balthica* (L.) from Chesapeake Bay. *J. Natl. Cancer Inst.* 52: 1739–1749; **Farley, C. A.** 1976. Proliferative disorders in bivalve mollusks. *Mar. Fish. Rev.* 38(10): 30–33.
41. **Brown, R. S., R. E. Wolke, S. B. Saila, and C. W. Brown.** 1977. Prevalence of neoplasia in 10 New England populations of the soft-shell clam *(Mya arenaria). Ann. N.Y. Acad. Sci.* 298: 522–534; **Yevich, P. P. and C. A. Barszcz.** 1977. Neoplasia in soft-shell clams *(Mya arenaria)* collected from oil-impacted sites. *Ann. N.Y. Acad. Sci.* 298: 409–426.
42. **Reinisch, C. L., A. M. Charles, and A. M. Stone.** 1984. Epizootic neoplasia in soft shell clams collected from New Bedford Harbor. *Hazard. Waste* 1: 73–81; **Farley, C. A., S. V. Otto, and C. L. Reinisch.** 1986. New occurrence of epizootic sarcoma in Chesapeake Bay soft shell clams *(Mya arenaria). Fish. Bull.* 84: 851–857.
43. **Young, J. S. and J. B. Pearce.** 1975. Shell disease in crabs and lobsters from New York Bight. *Mar. Pollut. Bull.* 6: 101–105.
44. **Young, J. S.** 1973. A marine kill in New Jersey coastal waters. *Mar. Pollut. Bull.* 4: 70; **Pearce, J. B.** 1972. The effects of solid waste disposal on benthic communities in the New York Bight, pp. 404–411. in Ruivo, M. (Ed.), *Marine Pollution and Sea Life*. Fish. News Ltd., London.
45. **Thomas, J. H.** 1954. The oxygen uptake of the lobster (*Homarus vulgaris* Edw.). *J. Exp. Biol.* 31: 228–251.
46. **Sawyer, T. K., S. A. MacLean, J. E. Bodammer, and B. A. Harke.** 1979. Gross and microscopical observations on gills of rock crabs *(Cancer irroratus)* and lobsters *(Homarus americanus)* from nearshore waters of the eastern United States, pp. 68–91. in Lewis, D.H. and J.K. Leong (Eds.), *Proceedings of the Second Biennial Crustacean Health Workshop*. Texas A&M University, Sea Grant Publ. No. 79–114; **Sawyer, T. K., E. J. Lewis, M. E. Galasso, J. J. Ziskowski, A. L. Pacheco, and S. W. Gorski.** 1985. Gill blackening and fouling in the rock crab, *Cancer irroratus,* as an indicator of coastal pollution, pp. 113–129. in Ketchum, B., J. Capuzzo, W. Butt, I. Duedall, P. K. Park, and D. Kester (Eds.), *Wastes in the Ocean*. Vol. 6. *Near-shore Waste Disposal*. John Wiley and Sons, New York.
47. **Estrella, B. T.** 1984. Black gill and shell disease in American lobster *(Homarus americanus)* as indicators of pollution in Massachusetts Bay and Buzzards Bay, Massachusetts. Mass. Dept. Fish. Wildl. Rec. Veh., Div. Mar. Fish. Publ. No. 14049–19–125–5-85-C.R. 17 pp.
48. **Krebs, C. T. and K. A. Burns.** 1977. Long term effects of an oil spill on populations of the salt-marsh crab *Uca pugnax*. *Science* 197: 484–487.

9 Effects of Pollution on Aquaculture

****

OYSTER CULTURE IN HIROSHIMA BAY

The late afternoon All-Nippon Airlines plane from Tokyo made a long gradual descent over the beautiful coastline of the Seto Inland Sea toward the Hiroshima airport. The year was 1966, and the coastal waters beneath the plane were crowded with the various structures vital to an expanding aquaculture technology — thousands of long rows of fence-like oyster spat collectors stretching seaward through the broad intertidal zone; hundreds of float-buoyed bamboo rafts in every protected cove, supporting dangling ropes on which young oysters were suspended; and, on shore, a great ferment of activity — building and repairing small boats, stringing shells for spat collection, preparing market-sized oysters for transport, and constructing new bamboo rafts.

At a meeting the next morning with members of the fishermen's cooperative and the staff of the national fisheries research laboratory located outside the city, the story of a successful and innovative oyster production system was told, and a profusion of meticulously drawn graphs illustrated ever-increasing yields, with projections of future growth. The meeting concluded with an elaborate lunch consisting entirely of products of aquaculture — most of them beautifully presented (and raw, of course). Oysters and shrimp dominated the menu.

Those of us from other parts of the world were genuinely impressed with the scale of the aquaculture operations, so much so that we neglected to ask our Japanese hosts some small niggling questions, such as the possibility of future problems with self-pollution and siltation from the masses of rafts and other inshore structures — or the possible impacts on oyster reproduction and growth of increasing industrial pollution from the nearby city of Hiroshima itself — or the possibility of organic overenrichment (eutrophication) of the Inland Sea, with consequent dangers from toxic algal blooms. We all flew away the next day, with these small questions unasked.

And so it was that when I had occasion to revisit Hiroshima two decades later, I looked again for all that aquaculture activity and all the physical evidence of a booming oyster industry, as the plane made its descent. But something was clearly wrong. The city, like so many others, had expanded enormously in the intervening time, and the shoreline was now thickly populated for as far as one could see. Yes, there were still a few oyster rafts here and there, but little could be seen of the intense aquaculture activity that was so apparent 20 years earlier. Colleagues at the research laboratory, when later queried about the obvious changes, told of major shifts in oyster culture away from the city, principally because of poor setting, poor survival, and poor growth in formerly productive waters, caused, they thought, by industrial chemical pollution and organic loading of the contained waters of the Inland Sea. They said that toxic algal blooms had occurred, and were becoming annual events in parts of the Inland Sea. The government scientists also reminded us that as long ago as 1970 they had warned that Japan's Inland Sea was becoming one of the most polluted areas in the world because of wastes from industrial plants. They did emphasize, however, an increasing government awareness and concern about the economic impacts of environmental degradation — an awareness intensified by demands from the powerful fishermen's cooperatives for drastic actions to halt the decline in water quality.

I flew away from Hiroshima the next day, as I had done once before so long ago — but this time with a little less optimism about human willingness to confront and to reduce coastal pollution problems — even in a country like Japan, where seafood is a major contributor of protein to the national diet, and a source of significant economic benefits.

Field Notes of a Pollution Watcher
(C. J. Sindermann, 1989)

****

INTRODUCTION

We have all heard overly dramatic but probably true pronouncements to the effect that "the edges of the sea cannot be used simultaneously as cesspools and as producers of food." Certainly for grossly polluted waters such as Boston Harbor and Raritan Bay, we can see effects such as the disappearance of oysters, clams, and certain fin fish from areas of former abundance, changes in abundance and diversity of food chain organisms, and spasmodic mortalities of some species when combinations of toxic chemicals, low oxygen levels, and high temperatures occur.

Effects of contaminants in other less grossly polluted bays, estuaries, and coastal waters, such as Biscayne Bay in Florida and certain parts of Delaware and Chesapeake Bays, are not as obvious. Marginal environmental conditions (toxic chemicals, low oxygen, diminished populations of certain forage organisms) may produce stresses on fish and shellfish that can result in poor condition, slow growth, greater susceptibility to infectious diseases, reduced survival of young, and slow attrition of survivors — all leading gradually, but less dramatically, to diminished wild stocks of species important to recreational or commercial fishermen and to reduced value of the degraded waters as locations for aquaculture facilities.

Additionally, a new pollution problem has become apparent as aquaculture operations have expanded in selected areas. This concerns *the aquaculture facility as a pollutant source*. Included would be feces, pseudofeces, and excretion products from the cultured animals, unconsumed food from fish-rearing pens, and chemicals used in disease control or as food additives. Nutrient loading of semi-enclosed coastal waters is a particular problem, since it can lead to eutrophication, toxic algal blooms, and hypoxia.

So, in examining pollution problems in coastal open-system aquaculture, three principal areas of concern can be identified: (1) *public health problems* produced by toxic industrial chemicals and microbial contaminants in domestic wastes; (2) *problems associated with negative effects of contaminants on reproduction, survival, and growth of cultured species;* and (3) *the environmental impacts of marine aquaculture* — modifications of the grow-out areas as a consequence of fish and shellfish culture. These problem areas will be explored in the following sections, keeping in mind the difficult question: "What real evidence do we have that industrial and domestic pollutants (and pollutants from fish farms) in estuaries and coastal waters of industrialized nations are incompatible with seafood production from open system (extensive) aquaculture"?

PUBLIC HEALTH PROBLEMS

Public health problems related to ocean pollution, of concern to aquaculture, can be categorized as those caused by *microbial contamination of food* and those caused by *chemical contamination of food*. Except for a few well-publicized chemical problems, such as Minamata Disease in Japan and high mercury levels in ocean-caught swordfish, much of the available information of public health significance concerns microbial contaminants. Public health problems related to *all* living marine resources were considered in Chapter 5, but it may be worthwhile to review here briefly those problems that seem most relevant to aquaculture.

Viral and Bacterial Contaminants

Microbial problems related to public health can be a major deterrent to open-system shellfish aquaculture. Viral diseases such as hepatitis have been shown repeatedly to be transmitted by ingestion of raw shellfish from polluted waters.[1] Viruses of terrestrial origin have been found experimentally to have variable, but in some instances surprisingly long, survival time in saline water.[2] Some limited information is available about their enhanced viability in sludges and in bottom sediments.

Viruses affecting humans, then, constitute a critical and ever-present problem for aquaculture operations in coastal and estuarine areas where even limited domestic pollution exists — and this includes most of the areas now used or planned for use in aquaculture. Possible viral contamination will be an important issue where treated sludges or other fecal degradation products are used for the enrichment of growing areas until large-scale inexpensive techniques are available that will absolutely guarantee viral destruction. The danger to public health in the U.S. from fecal contamination of growing areas in foreign countries that export aquaculture products to us has increased with our increased importation of seafood. Fortunately, the Food and Drug Administration has been aware of the problem, and, in addition to insisting on environmental data, such as coliform counts in export oyster-producing areas (in Japan, Korea, and other countries), the agency has even sent teams of experts to evaluate the risks of contamination. A large number of countries, but especially Korea, Taiwan, the Philippines, and now China, export aquaculture products to the U.S.; any contamination of important producing areas can seriously affect success in penetrating and holding positions in American markets. At the moment, a partial control measure is to import into the U.S. only processed products (canned or frozen), in which the danger of residual microbial contamination from foreign growing areas is minimized.

While the viruses constitute the most vexing public health problem in open-system aquaculture, pathogenic enteric bacteria form a continuing threat when raw or partially processed products are consumed by humans. Much attention has been paid to the role of marine vibrios (*Vibrio parahaemolyticus, V. cholerae,* and *V. vulnificus*) in outbreaks of human disease in the Orient, in Europe, and in the U.S. While vibrios are normal constituents of the inshore flora, their abundance may be increased facultatively by the organic enrichment of coastal and estuarine areas or by direct additions of pathogens with sewage. Other pollution-associated bacteria, such as *Clostridium, Salmonella,* and *Shigella,* must not be ignored, since a single outbreak of human disease that can be traced to any marine species may have a drastic impact on markets for *all* marine products, including those from aquaculture.

Very disturbing examples of the potential danger to humans from bacterial contamination of coastal waters and of cultivated marine animals have been discussed sporadically in the world press, beginning 20 years ago. Cholera outbreaks, caused by *V. cholerae* and traced to consumption of raw mussels from heavily polluted coastal waters, occurred in 1973 in two Italian cities,

Naples and Bari. This scourge from the Dark Ages has persisted in areas of extreme poverty and poor sanitation, and has had a resurgence in Asia and South America in this century. The causative bacterium is a normal inhabitant of brackish waters, but can survive for weeks in seawater and is accumulated by molluscan shellfish growing in polluted waters.

Whenever even one case of cholera occurs, the danger of shellfish contamination exists. A study in Portugal disclosed the presence of *V. cholerae* in 38% of 166 samples of molluscan shellfish taken in 1974 from the vicinity of Tavira, where a case of cholera was reported.[3] This report was a sequel to an earlier paper pointing out extensive pollution of shellfish growing areas on the southern (Algarve) coast of Portugal.[4] The *New York Times* reported in 1975 over 200 cases of cholera, with three deaths in Coimbra, Portugal. Health authorities attributed the outbreak to contaminated cockles, *Cerastoderma edule,* from the Mondego River estuary. Cholera has appeared sporadically in the United States in the past two decades, usually associated with the ingestion of raw or improperly processed seafood. Beginning in 1991, a major cholera outbreak occurred in South and Central America, with a principal focus in Peru. That outbreak was examined in Chapter 5.

Chemical Contaminants

Of the many chemicals that could occur in seafoods at levels toxic to humans, mercury has justly received the most attention. The horrors of Minamata Disease, caused by mercury contamination of fish and cultivated shellfish in a bay in Japan, were publicized over 30 years ago on television, in news magazines, and in books.[5] Hundreds died, and many others were permanently disabled after ingesting the contaminated seafood. Later, high mercury contaminant levels in swordfish served in 1971 to eliminate a developing fishery in the Gulf of Mexico and to cause a temporary ban on the sale of swordfish in the U.S. Increased surveillance of mercury and other heavy metals in all kinds of seafood lessens the likelihood of another Minamata incident, although whenever food is grown near industrial operations there is always a risk of chemical contamination, through negligence, deliberate actions, or accidental spills. Surveillance methods cannot easily be extended to include every localized aquaculture area, especially since some of the analytical methods are very time-consuming and the toxic levels for some contaminants are not known or are unstated.

The possible toxic effects of pesticides in foods have been evaluated and discussed on many occasions. Even though indiscriminate use of pesticides is coming under some measure of control in the U.S. and the Western world generally, their use in other parts of the globe is expanding. Because of their persistence in the environment, and their accumulation by successive levels of food chains, pesticides continue to be threats in near-shore ocean areas, including those devoted to marine aquaculture.

Certain petroleum derivatives are carcinogenic, and this fact has been used, particularly in the public press, to argue against offshore oil drilling and the establishment of offshore oil terminals. Points made are that oil is persistent for long periods in the marine environment, that oil hydrocarbons are concentrated by food chain organisms and transferred to humans when they consume seafood. It should be perfectly clear that there is at present no evidence for direct association of carcinogens in seawater, fish, shellfish, or bottom sediments with cancer in humans — that no case of human cancer has been unequivocally traced to chemical contaminants from the marine environment. However, this should not imply that there is no present or future danger from contaminated marine sources.

An additional public health problem of a chemical nature is that of biotoxins — several kinds of shellfish poisoning and ciguatera (fish poisoning) in particular. Seasonal toxicity to humans has characterized molluscan species from certain coastal areas; and some species of fish from particular locations in tropical and subtropical waters have long been known to be toxic. Changes in the distribution and intensity of toxicity have occurred in the past three decades, which may be related to human-induced modifications in coastal waters. European reports of paralytic shellfish poisoning (PSP) outbreaks seem related to very high nutrient levels in some coastal and harbor areas. Concerning ciguatera poisoning, one expert[6] has stated that "the field evidence...strongly suggests that under certain environmental conditions in tropical insular areas, pollutants may provide the necessary chemical constituents to trigger naturally occurring biotoxicity cycles such as ciguatera...." Fish poisoning in several forms is an important fisheries economic and public health problem in the Caribbean and in tropical Pacific islands. Possible pollution-mediated biotoxins must be a paramount consideration in planning future aquaculture sites, particularly (but not exclusively) in tropical and subtropical waters.

So, while there is little evidence of significant direct danger to human life in *existing* chemical contaminant levels in marine resource species (except for localized incidents related to gross pollution), there are instances (such as mercury in black marlin, large halibut, and swordfish) of levels of certain chemicals that are high enough to warrant attention, further study, and possibly controlled consumption. There is also a great need for much more scientific examination of possible long-term sublethal effects on humans of contaminants in seafood — especially that produced in aquaculture in marginally polluted coastal waters.

PROBLEMS ASSOCIATED WITH SURVIVAL, GROWTH, AND REPRODUCTION

Estuarine and coastal species, including those important to aquaculture, survive in a variable chemical, thermal, and biological environment. Effluvia from human populations can modify factors of that environment beyond levels tolerable to certain species, or can modify environmental factors enough to place continuing stress on the species, or can introduce new chemical factors with which the species has had no previous evolutionary experience. A study of the effects of such man-made changes on resource species is therefore a vital aspect of developing aquaculture technology. Principal problem areas can again be generally categorized as microbial and chemical, although physical changes in habitats could also be considered.

Bacterial and Other Diseases of Fish and Shellfish

The rich organic soups created in inshore areas by domestic sewage outfalls and sludge dumping can contribute to the expansion and modification of the normal marine and estuarine microbial flora. Wherever the organic content of water or sediments is increased, as it is in most aquaculture areas, there are opportunities for increases in populations of heterotrophic bacteria, such as vibrios, pseudomonads, and aeromonads. Concentrations of such bacteria may provide sufficient infection pressure on already stressed fish or shellfish so that disease and mortalities result. Some evidence for this sequence of events has been found in studies conducted in heavily polluted waters in which bacterial fin erosion characterized many species of fish.[7]

Indications of what may be a special microbial problem for aquaculture operations in heated effluents have been noted. Fossil fuel and nuclear electric generating plants produce significant amounts of heated water, which, for those installations located on estuaries or bays in temperate latitudes, has been widely publicized for its positive role in permitting year-round growth of aquaculture animals in thermally regulated warm water. Growth rates of species such as plaice, American lobsters, and oysters have been found to be significantly greater at elevated temperatures. There are, of course, mechanical problems created by unforeseen shutdowns in winter and the use of toxic antifouling chemicals in cooling tubes, but these are amenable to long-term solution. Evidence is accumulating, however, to suggest that the positive effects of thermal additions may be offset by increased mortality due to disease. A lethal viral disease of oysters held in heated discharge water in Maine has been described.[8] The disease, apparently existing in latent form in oysters growing at normal low environmental temperatures (12 to 18°C summer temperatures), seems to be enhanced in its effects on oysters at elevated temperatures (28 to 30°C).

Another viral disease, this one of cultured turbot in Scotland, produced mortalities in juveniles held in heated effluents, but was not lethal at usual environmental temperatures.[9] Yellowtail, cultured extensively in Japan, were found to suffer continuing high levels of disease-caused mortalities when held in heated effluents. Still another example of viral disease in fish held in heated effluents is lymphocystis disease of striped bass. High prevalences of the disease were found in fish overwintering in heated plumes of water. This viral disease is usually considered rare in striped bass, and its unusual abundance in a localized population may well be related to the abnormally high winter temperature regime in which the population exists. As with the oyster viral disease, high temperatures may promote survival or transfer of the pathogen or lower resistance of the host as a consequence of prolonged stress, resulting in grossly recognizable stages of infection.

Effects of Chemical Pollutants

Human-induced chemical changes in coastal/estuarine waters are of particular concern to aquaculture. Pesticides and herbicides are introduced in runoff from agricultural areas, heavy metals and PCBs in industrial effluents, and petroleum and its derivatives in polluted air as well as from off-loading facilities and accidental spills. Experimental evidence is accumulating slowly about the acute effects of high levels of such contaminants and about the probably more harmful chronic effects of lower levels of the chemicals. Since most of the global ocean has already been modified chemically to some extent by human activities, and since inshore waters (including those near present or potential aquaculture areas) have been and are being modified, it is important to know the extent of change and the tolerance of various life history stages of each aquaculture species to such changes. Of particular importance is the vulnerability of larval and early juvenile stages to cumulative environmental degradation, as well as the possible genetic effects on subsequent generations.

The extreme sensitivity of molluscan larvae to pollutants has been demonstrated and even used in the development of bioassay techniques for the detection of pollution.[10] Evidence of molluscan larval mortality after experimental exposure to contaminants has been reported from many laboratories and countries. Effects on embryos and larvae of oysters and clams have been and are being investigated in this country by several institutions; lethal effects and developmental inhibition at low concentrations of certain heavy metals are among the important findings to date.

The effects of pesticides on early life stages of cultured animals have long been a matter of concern. Severe growth retardation and the lethal effects of 52 commercial pesticides and herbicides on oyster and clam embryos and larvae were described more than two decades ago.[11] Growth rates of juvenile oysters were experimentally depressed by pesticides in other early studies, and residues of these compounds in oysters sampled from certain U.S. growing areas exceeded levels at which growth was retarded, indicating an *already existing* deleterious impact on oyster populations.[12] Furthermore, pesticide-contaminated diets fed experimentally to shrimp, crabs, and fish (croaker and pinfish) caused significant and rapid mortalities. The effects of pesticides on the reproduction of fish have been noted in many other studies, and it is quite likely that pesticide pollution causes significant increases in mortality at sensitive stages in estuarine fish and shellfish populations.

The implications of pesticide contamination in aquaculture areas are clear, but often overlooked or the consequences avoided. The sudden lethal intrusion of such chemicals from treated agricultural areas following heavy rains could cause mass mortalities, even eliminating entire marine crops, while lower levels could degrade otherwise excellent marine aquaculture areas to the status of "marginal" or "unproductive" for most species. This is a situation parallel to that in which "pristine" shellfish grow-out areas are contaminated with fecal bacteria, resulting in reclassification to "marginal" or "closed" status.

In addition to pesticides, other chlorinated hydrocarbons associated with industrial processes, including those commonly referred to as PCBs, are widely distributed in the marine environment, and have been the subject of countless studies. Extreme toxicity to juvenile shrimp has been demonstrated,[13] suggesting an added negative effect of pollutants on shrimp and possibly other crustacean aquaculture species.

Petroleum residues pose another chemical problem that could have significance to marine aquaculture. The risk of massive contamination of growing areas by accidental spills will always be present. In addition to such isolated catastrophes, the long-term contamination of coastal waters by airborne fallout, pleasure boat exhausts, sewage treatment plant discharges, and nonpoint pollutants carried seaward by riparian systems can affect the growth and quality of marine animals. Studies at the University of Rhode Island indicated that hydrocarbon pollutants caused a "stress syndrome" in clams that included reduction in growth rate, poor condition, shortened life span, and various physiological abnormalities.[14] Other studies disclosed that mullet took up petroleum hydrocarbons from coastal waters with very low levels of pollution.[15] Fatty infiltration of the liver and high lipid content of muscle characterized the affected fish. Still other investigations revealed that fish eggs and larvae were very sensitive to various crude oils. Mortality of eggs was proportional to the oil concentration, and larvae were more sensitive than embryos.[16]

Because bays and estuaries are particularly vulnerable to oil spills, the effects of petroleum on early life stages of shellfish have been of particular interest. In one study,[17] crude oil was toxic to lobster larvae at 100 ppm, with sublethal effects at concentrations as low as 1 ppm. In another study,[18] a significant decrease occurred in the fertilization rate of eggs of oysters and mussels exposed to crude oil. Sperm were particularly vulnerable, suggesting a possible pathway to reduced fecundity of bivalve populations in oil-polluted estuaries and embayments.

Adult clams can also be affected adversely by local oil spills and residual contamination of beds. An investigation in Maine[19] found that production from contaminated areas declined 20% in two years, while adjacent uncontaminated areas increased in yield nearly 250%. Clams moved to contaminated areas declined 65% in growth after transplantation, and survival was significantly poorer than in uncontaminated control areas.

These examples clearly demonstrate the effects of petroleum contamination on survival, reproduction, and growth of aquaculture species. Probably of more direct concern would be the tainting and rejection of aquaculture products as a result of oil spills or other types of deliberate petroleum discharges in coastal waters.

ENVIRONMENTAL INTERACTIONS OF MARINE AQUACULTURE

This story of pollution and aquaculture would be incomplete without considering quite a different perspective — that of *the environmental impacts of marine aquaculture — the role of aquaculture operations themselves as sources of pollution*. For logical consideration, this topic can be immediately subdivided into two major units: (1) *chemical modification* of coastal habitats and (2) *biological interactions* of aquaculture stocks and native populations (Table 3). *Both can be considered broadly as forms of habitat contamination.*

TABLE 3
Environmental Interactions (Impacts) of Coastal Aquaculture

Chemical Modifications	Biological Interactions of Aquaculture and Native Populations
• Chemicals used in disease and parasite control Dichlorvos: salmon lice Oxolinic acid: bacteria Oxytetracycline: bacteria • Nutrient loading — total N and P — from feces, urine, pseudofeces, unused food Leads to eutrophication, toxic blooms, hypoxia/reduced water quality, H_2S formation	• Introduced species: example, coho salmon in NE • Escapes of aquaculture species and interbreeding with natives; example, Atlantic salmon • Introduced macroalgae; example, *Undaria* • Introduced parasites and diseases; example, eelworm in Europe • Enhancement of parasite populations and impacts on native species; example, salmon parasites affecting sea trout in Ireland

CHEMICAL MODIFICATION OF COASTAL HABITATS

Aquaculture operations in coastal waters result in organic enrichment of the bottom sediments in the form of feces, pseudofeces (in the case of shellfish), and unused food (in the case of fish net pens and cages). This organic material, rich in nitrogen and phosphorus, may be resuspended, ingested by detritus feeders, or decomposed to simpler chemical forms. For the localized area, this process represents nutrient loading, and if the aquaculture operation is extensive, the affected area will be extensive. Possible effects include changes in species composition of the phytoplankton; an increase in primary production, sometimes resulting in algal blooms; and the development of hypoxia, anoxia, and hydrogen sulfide in and near the bottom, creating conditions inimical to marine life, including the fish being cultured.

Some researchers distinguish two aspects of the organic enrichment problem: (1) *organic site specific pollution* or waste-related pollution derived from lost food in net pen culture (up to 30% of the feed distributed in some instances); and (2) *eutrophication or nutrient enrichment* from excretions (ammonia and urea) and fecal matter. Feed formulations may contain up to 45% protein, so the waste portions are good sources of nitrogen. Lost feed may accumulate under net pens or in depositional areas; sediments become anoxic and hydrogen sulfide may be formed. Of the portion of the feed consumed by fish, Norwegian work with Atlantic salmon has indicated that 25% of the N input is converted into fish flesh, 15% is fecal matter, and 60% dissolves directly in seawater as excreted ammonia and urea.[20]

In coastal areas where aquaculture is a major industry — for example, in the inshore waters of the island of Shikoku in Japan — nutrient loading from net pen culture of yellowtail has produced the effects just described and has led to restrictions on the numbers of culture units that can be placed in each bay.[21] This has served as a means of limiting the negative environmental changes, especially the greater frequency and intensity of algal blooms that have resulted in mass mortalities.

Other kinds of chemicals enter the environment near aquaculture facilities. Notable are those used in disease and parasite control in fish culture. Antibacterial agents such as oxolinic acid, tetracycline, and some of the sulfas may be added to feed formulas to treat bacterial diseases, and even pesticides, such as nuvon or dichlorvos, may be used in dips to control the parasitic copepods that have emerged as a significant problem in salmon net pen culture in Norway, Scotland, and Canada. Some of the control agents, such as dichlorvos, are applied in large quantities and can be toxic to fish and shellfish larvae in areas affected by the application.[22] Also, the antibiotics released in the environment may promote resistant strains of endemic microorganisms.

Still other chemicals may enter coastal waters as a consequence of aquaculture activities. The list includes disinfectants, antifouling paints, and feed additives such as antioxidants, some of which may be toxic to native marine species.

Biological Interactions of Aquaculture Stocks and Native Populations

The presence of a large aquaculture facility in a coastal area — whether it be a salmon net pen farm, a dense concentration of yellowtail cages, or a series of mussel rafts — is certain to result in a number of biological interactions with native species. One activity that has caused much concern is the importation and culture of nonindigenous species (such as coho salmon in France), with the possibility of introducing exotic pathogens that may affect native populations of related species.[23] The natives might not have resistance to the introduced disease agent, and epizootics could occur. Just such a sequence of events occurred recently in European eels. Beginning in 1980, Japanese eels were imported to Germany. They carried a large and destructive parasitic worm, a nematode, to which European eels were very susceptible. The worms have spread in the past decade to almost all of Europe and have caused severe losses in eel production.[24]

Other than the possible introduction of exotic diseases with species imported for aquaculture purposes, there is the larger matter of escapes (from net pens, for example), which may lead to the development of self-sustaining populations of the import. The introduced species may prey upon young of native species, may compete with natives for food or spawning sites, or may modify habitats to the detriment of native stocks. Additionally, the introduced species may interbreed with closely related native species, possibly reducing the survival potential and adaptation level of the natives.[25]

Even if the aquacultured species is indigenous, there may be increasing distinctions between cultured and wild stocks that must be preserved. So, for example, genetic differences between wild and cultivated Atlantic salmon in Norway have been recognized, and concern has been expressed about the survival of escapees from net pens when they intermix with wild stocks and about the possible reduction of adaptation in offspring when genetic intermixing occurs.

One ramification of the disease threat to native species from large-scale coastal aquaculture seems to be emerging at present in several European countries, where stocks of sea trout, *Salmo trutta* have declined or collapsed since 1989. A relationship has been proposed with the simultaneous increase in farming of Atlantic salmon, *Salmo salar,* and a coincident enhancement of populations of parasitic copepods that can attack and affect sea trout as well as salmon. These are not *introduced* parasites, but their enormous increase in numbers because of the superabundance of salmon hosts in net pen culture could well have a spin-off effect on wild sea trout populations in the vicinity of the farms.

And finally, in this section on biological interactions of aquaculture and native species, we should consider the aquatic plants that have commercial value, usually as food. Marine macroalgae are harvested for food in surprising tonnages in many parts of the world, and species such as *Undaria* have long formed the basis of large-scale aquaculture in Japan. Other species, such as *Laminaria,* are sources of industrial products. Some of these economically valuable algae have been introduced to foreign coastal environments, where they may compete successfully and even replace native species, thereby altering that environment drastically while providing an income source. An excellent example has been the introduction and subsequent deliberate seeding of the alga *Undaria pinnatifida* on the coasts of France. First introduced accidentally with Japanese oysters on the Mediterranean coast, it prospered and was then transplanted deliberately to aquaculture facilities on the French Atlantic coast, where it also seems to be thriving.[26] It is reproducing in the new habitat and may well spread to coastal waters of other countries (with or without their approval).

In summary, then, the examples cited in this section on biological "interactions" of cultured and wild populations really describe different kinds of "biological pollution" in which aquaculture exerts a negative influence on native species and their habitats.

CONCLUSIONS

The best and the worst uses of coastal/estuarine waters of the planet, from a human perspective, are typified, respectively, by aquaculture and toxic waste disposal. The two activities are absolutely inimical, representing totally opposite assignment of priorities for the role of near-shore waters in the human economy. Existing and potential levels of pollution impose two severe difficulties related to living resources, especially to animals grown in open-system (extensive) aquaculture. One is danger to public health as a consequence of microbial or chemical contamination of the products of aquaculture. The other is danger to the survival, reproduction, and growth of cultured species resulting from microbial pathogens or from toxic levels of contaminants in the environment. Evidence has accumulated to demonstrate the reality of both concerns; examples exist of human disease outbreaks traceable to contaminated shellfish, and lethal or sublethal effects of contaminants on resource species have been demonstrated. With the expansion of marine aquaculture in some parts of the world, a new pollution problem has emerged: chemical and biological modification of the aquaculture sites and adjacent waters by the culture operations themselves.

Inhabitants of coastal and estuarine waters, where much of the fisheries and all of marine open-system aquaculture take place, are highly vulnerable to pollutants of all categories flushed down from the land, dumped from ships and boats, or released into the atmosphere. Eggs, larvae, juveniles, and adults of cultured species are all vulnerable to acute and chronic chemical poisoning, as well as to other forms of environmental stress.

Long-term persistence, synergistic effects, and bioaccumulation of pollutants make existing maximum allowable limits for specific contaminants questionable, especially since such levels are usually based on acute toxicity tests rather than long-term chronic tests. The survival, growth, and reproduction of resource animals in low concentrations of pollutants must be determined, since such concentrations are and will continue to be realities in most coastal and estuarine areas used for aquaculture. An unsatisfactory alternative would be to retreat from even marginally contaminated waters to a system of land-based aquaculture.

A few of the specific pollution problems that have emerged recently and that persist today are:

- Grow-out areas may be subject to aerial contamination.
- Fish eggs may contain residues of pesticides, and such eggs may have high hatching mortality or may produce abnormal young.
- Commercial feeds for young fish may contain pesticide residues that can contaminate grow-out areas and affect the survival and growth of nontarget species.
- Production may be affected by drastic increases in the number and severity of toxic algal blooms, such as those that occurred in China's coastal waters in 1993 and helped to reduce shrimp aquaculture production from 140,000 tons (in 1992) to an estimated 40,000 tons in 1993.

One of the critical present needs in marine aquaculture is a definition of water quality and other characteristics that will permit optimum reproduction, survival, and growth of each species selected for culture. At present, such a definition usually receives attention only after substantial investment has been made in coastal facilities or after problems of mortality or poor reproduction and growth have emerged. Once an aquaculture area with suitable water quality has been established, the strongest possible legal steps should be taken to prevent any degradation, even in surrounding zones. Damage from industrial pollution to aquaculture areas in Japan — a nation heavily committed to food production from the sea — has been reported for the past two decades, and those findings serve as a clear warning to the rest of the world of danger in the absence of adequate environmental protection. It seems clear that viable aquaculture operations must depend on management, not only of coastal/estuarine waters, but also of terrestrial sources of contamination.

Accepting, however unwillingly, the premise that coastal and estuarine pollution will decrease only very gradually, if at all, research should continue on low-cost mass techniques for the removal of contaminants from flow-through seawater systems. Ultraviolet, ozone, and resin columns are some approaches that have been tried with success for particular classes of contaminants, but methods presently available add significantly to production costs in aquaculture.

International standards must be developed and implemented — and a legal structure created — to regulate discharge of pollutants into marine waters. The first tangible step in this direction was taken more than 20 years ago, in June 1972, when the United Nations (U.N.) Stockholm Conference on the Human Environment adopted the principle that "States shall take all possible steps to prevent pollution of the seas by substances that are liable to create hazards to human health, to harm living resources and marine life, to damage amenities or to interfere with other legitimate uses of the sea." The second tangible step was taken in November of that same year, when 79 nations signed an international convention banning the dumping of oil, mercury, lead, zinc, arsenic and cadmium compounds, and highly radioactive wastes, and requiring permits for all other materials. Unfortunately, the recent (June 1992) U.N. Conference on the Environment held in Rio de Janeiro did not indicate, beyond pious cautionary statements, significant progress in global action to reduce ocean pollution. We still have a great distance to travel insofar as national commitments to enforcement are concerned, and the political maze is bewildering, but at least some tentative steps have been taken in a climate of worldwide concern about the state of the oceans and their ability to produce food. A guiding philosophy in all of aquaculture development should be that *because of an overriding need for clean water, aquaculture must be the single strongest force for marine environmental protection and pollution abatement in coastal/estuarine waters.*

REFERENCES

1. **Mason, J. O. and W. R. McLean.** 1962. Infectious hepatitis traced to the consumption of raw oysters. An epidemiologic study. *Am. J. Hyg.* 75: 90; **Richards, G.** 1985. Outbreaks of shellfish-associated enteric virus illness in the United States: requisite for development of viral guidelines. *J. Food Prot.* 48: 815–823.
2. **Metcalf, T. G. and W. C. Stiles.** 1966. Survival of enteric viruses in estuary waters and shellfish, pp. 439–447. in Berg, G. (Ed.), *Transmission of Viruses by the Water Route*. Wiley-Interscience, New York; **Canzonier, W. J.** 1971. Accumulation and elimination of coliphage S-13 by the hard clam, *Mercenaria mercenaria*. *Appl. Microbiol.* 21: 1024–1031.
3. **Ferreira, P. S. and R. A. Cachola.** 1975. *Vibrio cholerae* El Tor in shellfish beds of the south coast of Portugal. *Int. Counc. Explor. Sea,* Doc. C.M.1975/K:18, 7 pp.
4. **Cachola, R. and M. C. Nunes.** 1974. Quelques aspects de la pollution bactériologique des centres producteurs de mollusques de l'Algarve (1963–1972). *Bol. Inf., Inst. Biol. Marit.* 13, 12 pp.
5. **Harada, M.** 1972. *Minamata Disease*. [In Japanese.] Iwanami Shoten, Tokyo, 274 pp.; **Huddle, N. and M. Reich.** 1975. *Island of Dreams: Environmental Crisis in Japan*. Autumn Press, New York, 225 pp.; **Smith, W. E. and A. M. Smith.** 1975. *Minamata*. Holt Rinehart and Winston, New York, 220 pp.
6. **Halstead, B. W.** 1971. Toxicity of marine organisms caused by pollutants. *FAO Tech. Conf. Mar. Pollut.,* Doc. MP/70-/R-6, 21 pp.
7. **Mahoney, J.** 1970. Special microbiological problems in estuarine and coastal areas, pp. 177–178. in *Progress in Sport Fishery Research 1969*. U.S. Department of the Interior, Fish and Wildlife Service, Bur. Sport Fish. Wildl. Resour. Publ. No. 88; **Murchelano, R. A.** 1982. Some pollution-associated diseases and abnormalities of marine fish and shellfish: a perspective for the New York Bight, pp. 327–346. in Mayer, G.F. (Ed.), *Ecological Stress and the New York Bight: Science and Management*. Estuarine Research Federation, Columbia, SC.
8. **Farley, C. A., W. G. Banfield, G. Kasnick, Jr., and W. S. Foster.** 1972. Oyster herpes-type virus. *Science* 178: 759–760.

9. **Buchanan, J. S., R. H. Richards, C. Sommerville, and C. R. Madeley.** 1978. A herpes-type virus from turbot (*Scophthalmus maximus* L.). *Vet. Rec.* 102: 527–528.
10. **Woelke, C. E.** 1967. Measurement of water quality with the Pacific oyster embryo bioassay. Water Qual. Criteria, ASTM, STP416, pp. 112–120. American Society for Testing Materials, Philadelphia; **Woelke, C. E.** 1968. Application of shellfish bioassay results to the Puget Sound pulp mill pollution problem. *Northwest Sci.* 42: 125–133.
11. **Davis, H. C. and H. Hidu.** 1969. Effects of pesticides on embryonic development of clams and oysters and on survival and growth of larvae. *Fish. Bull. U.S.* 67: 393–404.
12. **Butler, P. A.** 1960. Effect of pesticides on oysters. *Proc. Natl. Shellfish. Assoc.* 51: 23; **Butler, P. A.** 1969a. The significance of DDT residues in estuarine fauna, pp. 205–220. in Miller, M.W. and G.G. Berg (Eds.), *Chemical Fallout.* Charles C Thomas, Springfield, IL; **Butler, P. A.** 1969b. Monitoring pesticide pollution. *Bioscience* 19: 889–891.
13. **Duke, T. W., J. I. Lowe, and A. J. Wilson, Jr.** 1970. A polychlorinated biphenyl (Aroclor 1254) in the water, sediment, and biota of Escambia Bay, Florida. *Bull. Environ. Contam. Toxicol.* 5: 171–180.
14. **Jeffries, H. P.** 1972. A stress syndrome in the hard clam, *Mercenaria mercenaria. J. Invertebr. Pathol.* 20: 242–251.
15. **Sidhu, G. S., G. L. Vale, J. Shipton, and K. E. Murray.** 1971. Nature and effects of a kerosene-like taint in mullet *(Mugil cephalus). FAO Fish. Rep.* No. 99: 143.
16. **Kuhnhold, W. W.** 1971. The influence of crude oils on fish fry. *FAO Fish. Rep.* No. 99: 157.
17. **Wells, P. G.** 1972. Influence of Venezuelan crude oil on lobster larvae. *Mar. Pollut. Bull.* 3: 105–106.
18. **Renzoni, A.** 1973. Influence of crude oil, derivatives and dispersants, on larvae. *Mar. Pollut. Bull.* 4(1): 9–13.
19. **Dow, R. L.** 1975. Reduced growth and survival of clams transplanted to an oil spill site. *Mar. Pollut. Bull.* 6: 124–125.
20. **Leffertstra, H.** 1992. Regulating effluents and wastes from aquaculture production: Norway, pp. 30–31. in Rosenthal, H. and V. Hilge (Eds.), *Proceedings of a Workshop on Fish Farm Effluents and Their Control in EC Countries,* Hamburg, Germany.
21. **Yasui, H. and E. Kobayashi.** 1991. Environmental management of the Seto Inland Sea. *Mar. Pollut. Bull.* 23: 485–488; **Kitamori, R.** 1992. Faunal and floral changes by pollution in the coastal waters of Japan. Second Int. Ocean Dev. Conf., Preprint Vol. 1, pp. 71–77.
22. ICES (International Council for the Exploration of the Sea). 1992. Report of the Working Group on Environmental Impacts of Mariculture. *Int. Counc. Explor. Sea,* Doc. C.M.1992/F:14, 95 pp.
23. **Sindermann, C. J.** 1993. Disease risks associated with importation of nonindigenous marine animals. *Mar. Fish. Rev.* 54(3): 1–10.
24. **Koie, M.** 1991. Swimbladder nematodes (*Anguillicola* spp.) and gill monogeneans (*Pseudodactylogyrus* ssp.) parasitic on the European eel *(Anguilla anguilla). J. Cons. Int. Explor. Mer* 47: 391–398.

25. NASCO (North Atlantic Salmon Conservation Organization). 1989. Report on Dublin meeting on genetic threats to wild stocks from salmon aquaculture. *NASCO Counc*. Pap. No. (89) 19, App. V, pp. 36–53.
26. **Floc'h, J. Y., R. Pajot and I. Wallentinus.** 1991. The Japanese brown alga *Undaria pinnatifida* on the coast of France and its possible establishment in European waters. *J. Cons. Int. Explor. Mer* 47: 379–390.

10 Global Changes in the Health and Condition of Marine Populations that May Be Associated with Human Activities

****

DIED AT SEA, OF UNKNOWN CAUSES

The early morning mist of late summer 1988 was disappearing from the beach at Manasquan, New Jersey; as it dissipated it revealed the bloated bodies of three sea mammals that had floated in with the tide during the night. They were young adult bottlenose dolphins — a species noted for a seemingly carefree existence in coastal waters and trained to be stars of tourist attractions at numerous oceanside aquaria. But they were, unaccountably, dead. They were also statistics, because during the summers of 1987 and 1988 more than 700 other members of their species had washed up on Atlantic shorelines — an unprecedented event.

What killed these graceful marine mammals? Was it a new viral disease that had reached epizootic levels in the population? Or had they encountered lethal levels of some industrial chemical, or some natural algal toxin? Or was some parasitic disease affecting equilibrium or respiration to the point where drowning could occur?

A task force of experts in marine diseases was assembled soon after the deaths began in 1987 — headed by a professor imported temporarily from a Canadian veterinary college. The group established field headquarters in 1987 and 1988 in oceanfront motels. The scientists also had technical support from their home laboratories.

Each group pursued various hypotheses as specimens were dissected and examined. News media representatives pushed daily for answers; tourists gathered in subdued clusters around dead animals on the beaches. After all the results of the various analyses — pathological, microbiological, and chemical — were completed, a tentative diagnosis was made and a final report prepared.[22] Deaths during the outbreak period seemed associated with high tissue levels of an algal biotoxin (brevetoxin) ingested with food. Weakened animals seemed less resistant to the microbial infections that were the immediate causes of death in most cases. Pathological effects may have been exacerbated also by exposure to industrial pollutants, which could create physiological stress and reduce immunocompetence. (Dolphins also died in large numbers — estimated to be in excess of 200 — in 1990 in the Gulf of Mexico).

Marine mammals — especially the dolphins — have some strange hold on the conscience of the human species. We seem to actually care about their well being. So it is natural for us, in rare moments of introspection, to examine our possible role in despoiling their habitats — to the degree that they can die in great numbers, as they did in 1987 and 1988.

From "Field Notes of a Pollution Watcher"
(C. J. Sindermann, 1992)

****

INTRODUCTION

A classic children's story depicts events stemming from an avian agitator, Chicken Little, spreading a false tale that the sky was falling. Her nonscientific interpretation of an acute environmental event (in my version she was hit on the head by a walnut) was incorrect. Having reached that conclusion, let me state that in this book on the effects of coastal pollution I want desperately to avoid being pigeonholed with a talking barnyard fowl or other environmental doomsayers — all those who predict impending catastrophes that either do not happen at all or become little more than whimpers at their predicted times.

A class of otherwise reputable scientists seems remarkably susceptible to this "doomspeaker's syndrome"; some are even subject to recurrences of the abnormality after they have been shown by the passage of time to be wrong. An outstanding example of recurrence would be the population biologist Paul Ehrlich, who in 1968 predicted a global famine within a decade. Now, almost a quarter-century later, he has again published essentially the same gloomy prediction, but with a 25-year slippage and with elaborate justifications for being wrong in his earlier prediction.

In close competition with Ehrlich for fallibility (but not for consistency) would be all those atmospheric scientists who predicted in the early 1970s that "the earth may be heading for another ice age." Are some of these the same scientists who now, *only two decades later,* are solemnly warning about "global warming" — or have all the members of the original cadre of ice people been replaced by their equally fallible students, now all grown up, ready to dispute their elders and full of different misconceptions?

The average newspaper-reading, television-watching American citizen is apt to get a little frustrated by such inconsistencies and even disillusioned with computer-assisted scientists as modern-day prophets. This could be expressed as increasing public concern about the validity of *any* conclusions reached by scientists, and maybe even as some reluctance to invest massive quantities of tax dollars to support gathering information that can lead to such erroneous conclusions.

So let me say clearly that I do not think that the oceans will become toxic for all life forms within the next 10 years. I do think, however, that despite the evolutionarily acquired resistance of many marine animals to deleterious changes in their environments (as examined in Chapter 2), local and large-scale events are occurring that suggest impaired health for these animals — at a level and scale that must be cause for concern. We have already dealt with localized acute and chronic problems of diseases in Chapters 3 and 4, but since at least some of the geographically more extensive events may be related also to human interference with natural processes in the sea — to pollution — it seems relevant to examine them more closely. This chapter does that, looking through one narrow window (that of marine pathology) at the separate phenomena of global occurrences of *ulcerations in fish, coral bleaching, mass mortalities of sea urchins, mortalities of marine mammals,* and *toxic algal blooms.*

ULCER DISEASES OF FISH

Ulcerations are highly visible abnormalities seen, usually in very low prevalences, in coastal/estuarine fish of many species. Lately though — within the past two decades — there have been numerous reports from various parts of the world of what can be described as *"epizootic ulcerative syndromes,"* in which the dominant gross pathology consists of superficial or penetrating ulcers and in which significant segments of local populations may be affected.[1]

Naturally, a search has been made for infectious agents — particularly viral, bacterial, or fungal — that may act as primary pathogens or secondary invaders. Such organisms have been found, and in some cases proposed as causative organisms responsible for the ulcerative condition. In many epizootics, though, the possible contributory or even dominant role of *environmental stressors,* such as sudden changes in temperature or salinity, extremes of these and other chemical and physical factors, increases in organic content of waters or industrial pollutants has been suggested, and some circumstantial evidence exists.

There is also information linking some ulcerative conditions with *microbial pathogens* such as *Vibrio anguillarum,* viruses, and oomycetous fungi, although even in these instances a role for environmental stressors has not been excluded. The possible roles of fungi in ulcerative syndromes are especially problematic. In some ulcerative syndromes (cod ulcer syndrome, spring ulcer disease of eels), fungi have not been reported; in others (red spot disease in Australasia, ulcerative diseases in Southeast Asia), fungi occur in advanced lesions and are considered secondary invaders; in still others (ulcerative mycosis of menhaden), fungi are considered primary pathogens. It seems reasonable that the usual sequence of events may be environmental stress, followed by primary microbial infection, followed in some cases by deep mycotic invasion, causing penetrating ulcers.

Recent increases in the frequency of reported occurrences of epizootic ulcerative syndromes in widely separated geographic areas invite the speculation that fish may be reacting to subtle environmental changes, effective over broad areas, resulting in increases in stress-related responses — one of which is the appearance of ulcerations. The nature of such environmental changes is not clear, but it is tempting to suspect a relationship to human activities, when we consider long-term modifications like the widespread dissemination of chlorinated hydrocarbons or gradual increases in organic loading of coastal/estuarine waters, with concomitant algal blooms and anoxic episodes. It is unlikely that any single stressor would have universal effects, but the totality of effects, in different areas and in different species, could raise the level of visibility of such an obvious indicator as fish ulcers.

The epizootic ulcerative conditions described here suggest the existence of major ecological disturbances — and not merely an artificial assemblage of disparate localized phenomena. If we exclude those ulcerations clearly associated with specific pathogens, such as red disease of Japanese eels caused by *Vibrio anguillarum,* we are still left with a worldwide array of epizootic ulcerative syndromes whose etiology is still uncertain (described as "idiopathic," in the usually opaque but occasionally delightful jargon of the pathologist), and for which environmental stressors seem to lurk in the background.

Ulcerations of fish thus constitute a broadly distributed phenomenon that can be provoked by abnormal environments. Other analogous examples exist: widespread coral bleaching, mass mortalities of sea urchins, mortalities of marine mammals, and toxic algal blooms.

CORAL BLEACHING

The widespread bleaching and death of corals was summarized recently.[2] Proposed causes include epizootic disease[3] and a number of natural and man-induced environmental changes (effects of El Nino, hurricanes, herbicide pollution, increased sedimentation, temperature increases, changes in nutrient concentrations). Because of the worldwide distribution of reported coral reef deterioration, a multifactorial etiology must be postulated, as has been done with epizootic ulcerative syndromes of fish, but, just as with fish ulcers, various forms of man-made pollution and environmental modifications have been suspected to be partly responsible for the observed phenomena.

The mechanism of coral bleaching has been described dramatically by the scientific team of Best and Lucy Williams.[4] Bleaching is caused by the loss of symbiotic single-celled algae (zooxanthellae) that normally colonize the tissues of corals. Different forms of stress (including temperature) can induce the normally white to faintly pigmented polyps to expel the zooxanthellae, producing a white patch in an otherwise greenish-brown coral colony. Suggested stressors capable of inducing bleaching include unusually high temperatures, increased ultraviolet radiation possibly due to ozone depletion, activity of secondary pathogens following physical stress, or some combination of these factors.

An extensive bleaching event was reported from the Caribbean in 1987, affecting reefs at Puerto Rico, Mona Island, Dominican Republic, Haiti, Cuba, Cayman Islands, Turks and Caicos, Jamaica, United States and British Virgin Islands, Bahamas, and the Florida Keys.[4] The depth range of effects extended from the surface to 60 meters, and divers in Puerto Rico described clouds of shed zooxanthellae around the reefs just before the bleaching was noted. This bleaching event of 1987 had been preceded by two comparable episodes — in 1980 and 1983 — in the Caribbean.[5] The causes were though to include increased global temperatures in the 1980s, progressive deterioration of inshore regions, and El Niño-related environmental changes.

Several recent reports summarize worldwide occurrences of coral bleaching extending far beyond the Caribbean to most of the tropical seas, especially the eastern Pacific, Indonesia, French Polynesia, southern Japan, and Australia.[6] Among the significant conclusions and opinions in these documents are these:

- Extensive coral bleaching and subsequent mortalities occurred during 1982 and 1983 in the eastern Pacific, coincident with the remarkable El Nino event of that period. As much as 70% of certain reefs was affected. Warm water anomalies of as much as 4 to 6°C in excess of normal seawater temperatures were recorded.[7]
- Coral bleaching and subsequent mortalities were also noted in the western Pacific, where El Niño-related phenomena of drought and increased salinities, and prolonged periods of low sea levels and aerial reef exposure were suggested as stressors.
- Extensive coral reef bleaching occurred in 1986 in the Hawaiian Islands and at other locations in the Pacific Ocean.[8]
- Bleaching may occur regularly on some reefs without causing extensive mortality.[9] Only when stress is prolonged or intense is bleaching followed by mortality.

There is some information that corals in bleached areas can recover, although the process may take four or more years.[10] Many reports document coral reef deterioration in the Atlantic and Pacific Oceans caused in part by human activities (sewage pollution and eutrophication, industrial pollution, sedimentation) and in part by natural phenomena (hurricanes, population explosions of predators, and El Niño-associated environmental changes). Bleaching can occur in relatively pristine as well as degraded habitats, indicating that pollution is not the sole cause of the phenomenon and that other stressors — especially high temperatures — must be involved. Some investigators, notably Williams and Bunkley-Williams,[5] have concluded that repeated episodes of coral reef bleaching represent worldwide interrelated disturbances. Principal causes, in their opinion, are unusually high seawater temperatures, exacerbated by El Niño's warming effects, and general worldwide deterioration of coral reefs, which lowers the resilience of the hosts to abnormal stressors.

SEA URCHIN MORTALITIES

Mass mortalities of several species of sea urchins have been reported within the past two decades from widely separated locations in the Western hemisphere — Canada, California, and the Caribbean.[11] As is the case with ulcer diseases of fish and bleaching of corals, an infectious agent has been suspected in at least some of the outbreaks, but an environmental component may also be involved.

Green sea urchins, *Strongylocentrotus droebachiensis,* died in large numbers in the Canadian Maritimes in the early 1980s.[12] Mortalities began in 1980 and were estimated to be as high as 70% in some near-shore waters of southwestern Nova Scotia in 1982.[13] Record high seawater temperatures seemed to be associated with the die-off, and experimental studies indicated a direct temperature relationship. However, other experimental studies suggested that death may have resulted from activities of a pathogen rather than, or in addition to, thermal stress, and infections by an amoeboid protistan parasite, tentatively described as a *Labyrinthomyxa* sp. (later labeled *Paramoeba* sp.), were seen in dying individuals.[14] Gross signs of morbidity were striking: spines appeared drooping and disheveled and were lost progressively; tube feet were retracted; jaws were gaping; and the epidermis was invaded by black lesions. Mortalities were observed along 440 km of coastline, extending out to depths of at least 13 m. Anecdotal information from fishermen indicated a previous mass mortality of sea urchins in southwestern Nova Scotia in the early 1950s, also a period of relatively high sea-surface temperatures.

As mortalities of green sea urchins in northern waters declined in the mid-1980s, new reports of mass deaths — this time of the black sea urchin, *Diadema antillarum* — came from widespread points in the Caribbean. Mortalities were first observed on the coast of Panama in January 1983, and within one year much of the rest of the Caribbean and even Bermuda were involved.[15]

The path of mortalities seemed to follow ocean current patterns from Panama, suggesting the existence of a waterborne pathogen rather than a pollutant or abnormal water mass, although a specific pathogen was not associated definitively with the mass mortalities. Probably the most striking aspect of these mortalities, whatever their cause, was the geographic area covered — some 3.5 million square kilometers — probably the most widespread outbreak ever recorded for a marine invertebrate (with the possible exception of coral bleaching).

Mortalities of black sea urchins in the Caribbean in the period 1983 to 1984 reached 95% on some reefs and have continued at a reduced level ever since, but with a possible resurgence in 1991 among remaining urchin populations. As with most other marine mortalities, local or widespread, pollution and general degradation of inshore habitats have been suspected as being contributory, as have increased water temperatures.

Mass mortalities of sea urchins off Nova Scotia and in the Caribbean have been the most spectacular from the perspective of geographic areas involved, but other species in other areas have also been affected. The red sea urchin, *Strongylocentrotus franciscanus,* was reported to suffer localized mortalities on the central California coast in 1970 and again in 1976.[16] Disease signs included large areas of the test denuded of spines and epidermis, with deep penetrating lesions. A fungal pathogen was suspected but not demonstrated. In another geographic area, sea urchin *(Paracentrotus lividus)* populations on the French Mediterranean coast suffered a precipitous decline within a one-year period in 1980. As with other investigations, an epizootic was suspected, but no causative agent was recognized.[17]

The mass mortalities of sea urchins described here may be associated with a variety of infectious agents, but in no case has a causative organism been conclusively identified.[18] Morbid signs in different geographic areas and in different species are similar — especially surface lesions, disorientation, and loss of spines — but these are probably generalized indicators of infection. In most of the geographic areas affected by mass mortalities, abnormally high water temperatures may have been predisposing factors.

MORTALITIES OF MARINE MAMMALS

Mass deaths have occurred during the past 15 years among populations of the two major groups of marine mammals — pinnipeds (seals, sea lions, fur seals, and walrus) and cetaceans (dolphins, porpoises, and whales). Viral infections have been reported to be primary causes of death in harbor seals and striped dolphins, whereas algal toxins have been implicated in some mortalities of bottlenose dolphins and humpback whales on the Atlantic coast of North America. Many reports allude to, but do not demonstrate, a possible role for marine pollution, particularly in reducing immune responses of the animals.

Among the pinnipeds, seal mortalities have been especially dramatic — both on the northern Atlantic coast of the United States and in northern European waters. During 1979 to 1980, more than 500 harbor seals, *Phoca vitulina,* died in New England waters; an influenza virus, probably of avian origin, was identified as the primary cause of death.[19] Then in the spring and summer of 1988, massive numbers of dead harbor seals — some estimates exceed 20,000 — were washed ashore in northern Europe, especially Norway, Sweden, Denmark, Britain, and Germany. A newly recognized virus, called phocine distemper virus (PDV), was identified as the primary cause of the mortalities (the virus belongs to the morbillivirus group, which also contains the organisms responsible for distemper in dogs and measles in humans).

The "seal plague" in northern Europe in 1988 was extensive enough to cause drastic reductions in population size, ranging from 40 to 60% of existing regional stocks.[20] Disease effects declined drastically in 1989. One fascinating aspect of this outbreak was that it was preceded (in the autumn of 1987) by a similar disease among Baikal seals, *Phoca siberica,* in Lake Baikal in southern Siberia. Several thousand seals died, and the virus that was isolated was also a morbillivirus, but one more closely related to canine distemper virus than to the phocine virus that killed harbor seals in Europe in 1988.[21]

Mass deaths of some cetaceans are also a matter of record. Mortalities of bottlenose dolphins, *Tursiops truncatus,* and humpback whales, *Megaptera novaeangliae,* occurred on the Atlantic coast of the United States in 1987 and 1988. Dolphin mortalities began in the summer of 1987 in New Jersey waters and persisted along the Atlantic seaboard for 11 months. During that period, more than 700 dolphins died. The humpback whale mortality occurred in late 1987 off the Massachusetts coast and involved 14 animals killed in a little over one month. Detailed examinations and chemical analyses of tissues led to the conclusion that mortalities were associated with the presence in the animals of the algal neurotoxins brevetoxin and saxitoxin.[22] The source of the toxins was hypothesized to be food fish, especially menhaden and mackerel, and effects included reduced physiological fitness and thus increased susceptibility to secondary bacterial infections that were the immediate cause of death.[23]

Mass mortalities of dolphins have also occurred recently in the Mediterranean. In 1990 and 1991, over 1,000 striped dolphins, *Stenella coeruleoalba,* died in the western and central Mediterranean; and in 1992 striped dolphins died in large numbers in the eastern Mediterranean. Unlike the toxin-induced deaths of dolphins in North America, these events in the Mediterranean seemed to be the result of infections by different strains of the morbillivirus group — phocine distemper viruses — that had killed seals a few years earlier in northern European waters. The impacts of the viral epizootics may have been exacerbated by reduced immune competence (lower disease resistance) induced by high levels of PCBs and other contaminants in food and in the environment.

TOXIC ALGAL BLOOMS

Population explosions of planktonic unicellular algae — so-called "algal blooms" or "red tides" (even though many of them are not red) have been observed for centuries and have in some instances caused shellfish in areas such as Puget Sound and northern New England to become temporarily toxic to humans. Paralytic shellfish poisoning (PSP) is the best-known consequence of eating toxic bivalve molluscs, although several other types of poisoning exist and were described in Chapter 5. Some blooms result in toxicity to humans and/or other animals and some do not. Interest in toxic algal blooms is usually stimulated by the danger of toxic effects on humans, usually from eating contaminated shellfish, but, as was pointed out emphatically in a recent review,[24] fish and shellfish (and other marine animals) may be affected severely by some of the algal toxins.

During the past three decades, there has been a real, rather than just a perceived, increase in the frequency, severity, and geographic extent of toxic algal blooms — toxic for fish, shellfish, marine mammals, and humans.[25] As an example, sea herring, *Clupea harengus,* were killed in large numbers in the Bay of Fundy in the late 1970s by feeding on zooplankton (cladocerans and pteropods) that had fed on the dinoflagellate *Gonyaulax excavata,* a species best known for causing paralytic shellfish poisoning (PSP) in humans.[26] In a later experimental study, red sea bream, *Pagrus major,* larvae and juveniles were affected, and many died after feeding on plankton containing *Gonyaulax* toxins.[27] Farmed fish, especially Atlantic salmon, have been killed in large numbers by algal blooms. A recent outbreak occurred on the Norwegian coast in 1988, where a large bloom of *Chrysochromulina polylepsis* caused extensive mortalities in sea cages.[28] Widespread blooms of this toxin-producing species occurred in waters adjacent to North European countries during much of the decade of the 1980s, seriously affecting salmon aquaculture.[29]

Major destructive blooms have also occurred in many other parts of the world, with severe effects on aquaculture. Notable are extensive and recurrent blooms in parts of the Seto Inland Sea of Japan that have affected yellowtail and red sea bream production.[30] Other sporadic outbreaks have had impacts on mussel culture in Spain and Canada, and on bay scallop production in Long Island (NY) waters. Toxic blooms have become important enough as a global problem to result in the establishment by the United Nations of a newsletter entitled "Harmful Algae News" (published by the Intergovernmental Oceanographic Commission (IOC) of UNESCO). Almost predictably, a new scientific publication, *The Journal of Natural Toxins,* has also been established (in 1993). The U.S. National Statement on Harmful Algal Blooms, presented in 1992 to IOC's Panel on Harmful Algal Blooms, effectively summarizes the U.S. perspective and the seriousness of the problem in the following excerpts from that document:[31]

Marine biotoxins and harmful algae represent a significant and expanding threat to human health and fisheries resources throughout the U.S. We have experienced different toxic or harmful algal species, more frequent and larger outbreaks, different toxins, and a growing list of affected resources. In spite of this trend, the reasons for the increasing scale of the problem and understanding of the biological, physical, and chemical processes that regulate harmful algal blooms (HABs) remain elusive.

The nature of the problem has changed considerably over the last two decades in the U.S. Where formerly a few regions were affected in scattered locations, now virtually every coastal state is threatened, in many cases over large geographic areas and by more than one harmful or toxic algal species. There is a growing consensus in the scientific community that the number of harmful events in U.S. waters and the economic costs associated with them have increased dramatically over the last several decades. Amnesic shellfish poisoning (ASP), unknown before 1987, occurred on the west coast in 1991. The causative organism has been identified in Northeast and Southeast waters of the U.S. as well. Paralytic shellfish poisoning (PSP), at first thought to be a problem only in shellfish, has been found in mackerel, and the toxins were implicated in whale deaths. Prominent examples of HAB-related problems in the U.S. include:

- Recurrent PSP outbreaks now affect the states of Maine, New Hampshire, Massachusetts, Oregon, Washington, and Alaska. PSP problems constrain the development of a shellfish industry in Alaska. Offshore shellfish on Georges Bank became toxic for the first time in 1989 and have remained toxic ever since. Low levels of PSP have been found in Rhode Island, Connecticut, and New York. In 1987, 19 whales died from PSP toxin contained in mackerel they had consumed.
- Neurotoxic shellfish poisoning (NSP), traditionally a problem only in the state of Florida, caused closures of major shellfish harvesting areas in North Carolina and South Carolina in 1987 and 1988. Hundreds of Atlantic dolphin died in 1988, possibly due to brevetoxin, the toxin that causes NSP.
- A brown tide caused scallop and other shellfish mortalities in New York and Rhode Island in 1985. Another major brown tide occurred along the south Texas coast from 1990 through 1992.
- ASP toxin was detected in Nantucket scallops in 1990 and 1991. Toxic *Nitzschia* species have been identified in the Gulf of Mexico. Seabird mortalities in the state of California in 1991 have been linked to levels of the toxin found in the flesh of fish that were consumed. In 1991, the ASP toxin also occurred in the state of Washington, where contaminated clams and crabs caused human illness.
- In 1990, the first confirmed outbreak of diarrhetic shellfish poisoning (DSP) in North America occurred in Canada. Two more DSP outbreaks occurred in Canadian waters in 1992. Scattered, unconfirmed cases of DSP have been reported in the U.S., and the causative organisms have been positively identified in U.S. waters. If DSP proliferates in the U.S., as has PSP, the impact on the U.S. shellfish industry will be devastating.
- Mortalities of farmed salmon in the Pacific Northwest due to blooms of the diatom *Chaetoceros* and the chloromonad *Heterosigma* have been a serious detriment to the development of this new industry.

- The incidence of ciguatera poisoning, caused by toxins produced by a dinoflagellate associated with coral reefs, appears to be rising in Florida, the Caribbean, and the Pacific. These toxins pass up the food chain from herbivorous reef fish to larger carnivorous, commercially valuable finfish. Ciguatera traditionally was limited to tropical regions; however, modern improvements in refrigeration and transport have augmented commercialization of tropical reef fish and increased the frequency of this type of fish poisoning in temperate regions.

Evidence is accumulating slowly that human activities — especially nutrient enrichment of coastal/estuarine waters by agricultural runoff, sludge dumping, sewerage outfalls, and some constituents of industrial effluents — may enhance the likelihood of blooms. For some of the blooms, the algal species that were responsible were either unknown previously in the outbreak area or were reported only rarely, leading to postulations that the toxic organisms may have been imported in ships' ballast water or attached to introduced shellfish species. This is of course a form of biological pollution, and some limited evidence for both methods of transfer has been reported from studies in Ireland and Australia.[32]

Beginning in the late 1980s, a new perception of mechanisms of algal toxicity has emerged, with the description of so-called "phantom algae" — encysted forms that can be induced to bloom and produce toxin very quickly by the presence of fish in the vicinity.[33] The toxins are rapidly lethal to fish, as determined by experimental exposures,[34] and previously unexplained fish kills have now been thought to result from this kind of severe environmental interaction.

The danger exists that if algal blooms continue to increase in numbers and effects, some aquaculture ventures may be severely compromised or even eliminated from coastal/estuarine locations where blooms recur too often. Concern has also been expressed about examples of greater duration of toxicity in natural populations of bivalve molluscs and the increasing costs of monitoring toxicity levels for public health purposes.

From an ecosystem perspective, it is possible to envision major shifts in the dominance of algal populations, in which species of different sizes and with different nutrient requirements may supplant previous species assemblages. Such changes could affect feeding relationships of higher links in food webs, eventually even affecting the abundance and distribution of fish and shellfish. A good example of this was seen during the recent (1985 to 1987) bloom of *Aureococcus anophagefferens* in Long Island bays. Hard shell clams, *Mercenaria mercenaria,* which are filter feeders, showed evidence of starvation in the midst of plenty, probably because the bloom organisms were too small to be accepted as food and because shell valves remained closed when toxin was present.[35] Similarly, bay scallops, *Argopecten irradians,* showed a 76% reduction in adductor muscle weight compared to the previous year.[36]

This discussion of toxic algae illustrates the concept that *change* — chemical, physical, and biological — rather than *stability* is the norm in coastal/estuarine environments. Changes can be dramatic and sudden, such as those resulting from a hurricane or an algal bloom, or gradual, such as the slow spread over decades or even centuries of anoxic zones in eutrophic estuaries. Encounters with toxic algal blooms have undoubtedly been part of the evolutionary history of coastal/estuarine animal species, and survival mechanisms have been developed on three levels: (1) the individual, which adapts or dies, such as the clam that closes its valves tightly when it first senses a toxin; (2) the population, which, after generations of exposure to toxins, consists mostly of individuals that have developed defense mechanisms and have modified reproductive strategies to counter the effects of the toxins; and (3) the community, which is modified in its species composition and dominance by the selective pressures of the toxins (and by many other influences). What is different in the recent past is the frequency, severity, and widespread distribution of toxic blooms in coastal waters — combined with the likelihood that at least some of them are caused by nonindigenous organisms introduced by humans as consequences of their commercial practices. These events raise the ante for individuals and species in terms of survival in toxic environments, by pushing harder on the adaptive strategies and responses that enable their continued existence in temporarily disturbed habitats.

CONCLUSIONS

Here, then, are summaries of five recent large-scale ecological events, as measured by geographic area affected and potential population impacts — ulcerations in fish, coral bleaching, mass mortalities of sea urchins, mortalities of marine mammals, and toxic algal blooms. Each represents an amplification of events that may have been observed on a local scale previously, but each has recently assumed dimensions that warrant examination from a global perspective. The need for such an examination is especially urgent if even some of these events are indicators of negative impacts of humans on the biosphere.

It can certainly be argued that these episodes may be controlled by natural physical and biological factors that we have too little information about (such as the role of El Niño in coral bleaching). It can also be argued that human intrusions into marine habitats are ubiquitous and are especially demonstrable in the coastal waters where the abnormalities and mortalities occur.

Reviewers of these events enter a dimly lit zone where available data lead to the formation of hypothetical scenarios to explain them. The conceptual leap to global interpretations of what may be only artificial aggregates of local phenomena is one that some scientists can accept, but others may reject.

Since my chosen specialty is marine pathology, it is hard for me to reject the information coming from worldwide reports of epizootic ulcerative diseases of fish, fresh water and marine. Obviously the etiology of lesions may vary from location to location and from country to country, but the reality of this stress-related abnormality cannot be denied. It is ubiquitous, and it may well be one of the best methods of communication on environmental degradation matters that we humans have with the bony fishes. I am equally convinced that the marine mammals have important messages for us about the deteriorating state of their chemical environment and the contamination of their food sources by industrial pollutants. Mass deaths of seals and dolphins in many parts of the world seem to be — sadly — their most effective method of reaching us.

REFERENCES

1. **Sindermann, C. J.** 1988. Epizootic ulcerative syndromes in coastal/estuarine fish. U.S. Department of Commerce, Natl. Mar. Fish. Serv., Woods Hole, MA. NOAA Tech. Memo. NMFS-F/NEC-54, 37 pp.
2. **Brown, B. E.** 1987. Worldwide death of corals — natural cyclical events or man-made pollution? *Mar. Pollut. Bull.* 18: 9–13.
3. **Gladfelter, W. B.** 1982. White band disease in *Acropora palmata:* implications for the structure and growth of shallow reefs. *Bull. Mar. Sci.* 32: 639–643; **Antonius, A.** 1985. Coral diseases in the Indo-Pacific. A first record. *Mar. Ecol.* 6: 197–218; **Glynn, P. W., E. C. Peters, and L. Muscatine.** 1985. Coral tissue microstructure and necrosis: relation to catastrophic coral mortality in Panama. *Dis. Aquat. Org.* 1: 29–37.
4. **Williams, L. B. and E. H. Williams, Jr.** 1988. Coral reef "bleaching" peril reported. *Oceanus* 30: 69–75.
5. **Williams, E. H., Jr. and L. Bunkley-Williams.** 1991. The worldwide coral reef bleaching cycle and related sources of coral mortality. Smithsonian Inst., *Atoll Res. Bull.,* 62 pp.
6. **Glynn, P. W.** 1984. Widespread coral mortality and the 1982–1983 El Nino warming event. *Environ. Conserv.* 10: 149–154; **Brown, B. E.** 1987. Worldwide death of corals — natural cyclical events or man-made pollution? *Mar. Pollut. Bull.* 18: 9–13; **Williams, E. H., Jr. and L. Bunkley-Williams.** 1991. The worldwide coral reef bleaching cycle and related sources of coral mortality. Smithsonian Inst., *Atoll Res. Bull.,* 62 pp.
7. **Glynn, P. W.** 1984. Widespread coral mortality and the 1982–1983 El Nino warming event. *Environ. Conserv.* 11: 133–146.
8. **Tsuchiya, M., K. Yanagiya, and M. Nishihara.** 1987. Mass mortality of the sea urchin *Echinometra mathaei* (Blainville) caused by high water temperature on the reef flats in Okinawa, Japan. *Galaxea* 6: 375–385; **Williams, E. H., Jr. and L. Bunkley-Williams.** 1991. The worldwide coral reef bleaching cycle and related sources of coral mortality. Smithsonian Inst., *Atoll Res. Bull.,* 62 pp.
9. **Brown, B. E.** 1987. Worldwide death of corals — natural cyclical events or man-made pollution? *Mar. Pollut. Bull.* 18: 9–13.
10. **Suharsono.** 1988. Monitoring coral reefs to assess the effects of seawater warming in 1982–1983 at Pari Island Complex, Thousand Island, Indonesia. *Abstr. 6th Int. Coral Reef Symp.,* Australia, p. 97; **Bythell, J.** 1989. Buck Island Reef long-term monitoring program. *West Indies Lab., Newsl.,* Winter 1988–89, St. Croix, USVI, p. 7.
11. **Pearse, J. S., D. P. Costa, M. B. Yellin, and C. R. Agegian.** 1977. Localized mass mortality of red sea urchin, *Strongylocentrotus franciscanus,* near Santa Cruz, California. *Fish. Bull. U.S.* 75: 645–648; **Scheibling, R. E. and R. L. Stephenson.** 1984. Mass mortality of *Strongylocentrotus droebachiensis*

(Echinodermata: Echinoidea) off Nova Scotia, Canada. *Mar. Biol.* 78: 153–164; **Lessios, H. A., D. R. Robertson, and J. D. Cubit.** 1984. Spread of *Diadema* mass mortality through the Caribbean. *Science* 226: 335–337.

12. **Miller, R. J. and A. G. Colodey.** 1983. Widespread mass mortalities of the green sea urchin in Nova Scotia, Canada. *Mar. Biol.* 73: 263–267.
13. **Scheibling, R. E. and R. L. Stephenson.** 1984. Mass mortality of *Strongylocentrotus droebachiensis* (Echinodermata: Echinoidea) off Nova Scotia, Canada. *Mar. Biol.* 78: 153–164.
14. **Li, M. F., J. W. Cornick, and R. J. Miller.** 1982. Studies of recent mortalities of the sea urchin *Strongylocentrotus droebachiensis* in Nova Scotia. *Int. Counc. Explor. Sea,* Doc. C.M./L:46; **Jones, G. M., A. J. Hebda, R. E. Scheibling, and R. J. Miller.** 1985. Histopathology of the disease causing mass mortality of sea urchins *(Strongylocentrotus droebachiensis)* in Nova Scotia. *J. Invertebr. Pathol.* 45: 260–271.
15. **Lessios, H. A., J. D. Cubit, D. R. Robertson, M. J. Shulman, M. R. Parker, S. D. Garrity, and S. C. Levings.** 1984. Mass mortality of *Diadema antillarum* on the Caribbean coast of Panama. *Coral Reefs* 3: 173–182; **Lessios, H. A.** 1988. Mass mortality of *Diadema antillarum* in the Caribbean: what have we learned? *Annu. Rev. Ecol. Syst.* 19: 371–393.
16. **Johnson, P. T.** 1971. Studies on diseased urchins from Pt. Loma, pp. 82–90. in North, W.J. (Ed.), *Kelp habitat improvement project, annual report 1970–1971.* California Institute of Technology, Pasadena; **Pearse, J. S., D. P. Costa, M. B. Yellin, and C. R. Agegian.** 1977. Localized mass mortality of red sea urchin, *Strongylocentrotus franciscanus,* near Santa Cruz, California. *Fish. Bull. U.S.* 75: 645–648.
17. **Boudouresque, C. F., H. Nedelec, and S. A. Shepherd.** 1981. The decline of a population of the sea urchin *Paracentrotus lividus* in the Bay of Port-Cros (Var. France). *Rapp. P.-V. Réun. Comm. Int. Explor. Sci. Mer Méditerr.* 27: 223–234; **Höbaus, E., L. Fenaux, and M. Hignette.** 1981. Premières observations sur les lesions provoquées par une maladie affectant le test des oursins en Méditerranée occidentale. *Rapp. P.-V. Réun. Comm. Int. Explor. Sci. Mer Méditerr.* 27: 221–222.
18. **Scheibling, R. E.** 1984. Echinoids, epizootics and ecological stability in the rocky subtidal off Nova Scotia, Canada. *Helgol. Meeresunters.* 37: 233–242.
19. **Geraci, J. R., D. J. Aubin, I. K. Barker, R. G. Webster, V. S. Hinshaw, W. J. Bean, H. L. Ruhnke, J. H. Prescott, G. Early, A. S. Baker, S. Madoff, and R. T. Schooley.** 1982. Mass mortality of harbor seals: pneumonia associated with influenza A virus. *Science* 215: 1129–1131.
20. **Tougaard, S.** 1989. Monitoring harbour seal *(Phoca vitulina)* in the Danish Wadden Sea. *Helgol. Meeresunters.* 43: 347–356; **Harkönen, T. and M.-P. Heide-Jorgensen.** 1990. Short-term effects of the mass dying of harbour seals in the Kattegat-Skagerrak area during 1988. *Z. Säugetierkde* 55: 233–238; ICES. 1990. Report of the joint meeting of the working group on Baltic seals and the study group on the effects of contaminants on marine mammals. *Int. Counc. Explor. Sea,* Doc. C.M.1990/N:14.
21. **Grachev, M. A., V. P. Kumarev, L. V. Mamaev, V. L. Zorin, L. V. Baranova, N. N. Denikina, S. I. Belikov, E. A. Petrov, V. S. Kolesnik, R. S. Kolesnik, V. M. Dorofeev, A. M. Beim, V. N. Kudelin, F. G. Nagieva, and V. N. Sidorov.** 1989. Distemper virus in Baikal seals. *Nature* (London) 338: 209.

22. **Geraci, J. R.** 1989. Clinical investigation of the 1987–88 mass mortality of bottlenose dolphins along the U.S. central and south Atlantic coast. Final Rep., Natl. Mar. Fish Serv. and U.S. Navy, Off. Naval Res. Mar. Mammal Comm., 63 pp.
23. **Anderson, D. M. and A. W. White (Eds.).** 1989. Toxic dinoflagellates and marine mammal mortalities. Tech Rep. No. CRC-89–6, Woods Hole Oceanographic Inst., Woods Hole, MA., 65 pp.
24. **Shumway, S. E.** 1989a. A review of the effects of algal blooms on shellfish and aquaculture. *Int. Counc. Explor. Sea,* Doc. C.M.1989/E:25, 38 pp.
25. **White, A. W.** 1988. Blooms of toxic algae worldwide: their effects on fish farming and shellfish resources, pp. 9–14. In *Proceedings of International Conference on Impact of Toxic Algae on Mariculture*. Aqua-Nor 87 International Fish Farming Exhibition, August 1987, Trondheim, Norway; **Smayda, T. J.** 1989. Primary production and the global epidemic of phytoplankton blooms in the sea: a linkage?, pp. 449–484. in Cosper, E.M., V.M. Bricelj, and E.J. Carpenter (Eds.), *Novel Phytoplankton Blooms. Coastal and Estuarine Studies No. 35*. Springer-Verlag, New York; **Smayda, T. J.** 1990. Novel and nuisance phytoplankton blooms in the sea: evidence for a global epidemic, pp. 29–40. in Granelli, E., B. Sundstrom, L. Edler, and D.M. Anderson (Eds.), *Toxic Marine Phytoplankton*. Elsevier, New York; **Shumway, S. E.** 1989b. Toxic algae: a serious threat to shellfish aquaculture. *World Aquacult.* 20: 65–74.
26. **White, A. W.** 1980. Recurrence of kills of Atlantic herring *(Clupea harengus harengus)* caused by dinoflagellate toxins transferred through herbivorous zooplankton. *Can. J. Fish. Aquat. Sci.* 37: 2262–2265.
27. **White, A. W., O. Fukuhara, and M. Anraku.** 1989. Mortality of fish larvae from eating toxic dinoflagellates or zooplankton containing dinoflagellate toxins, pp. 395–398. in Okaichi, T., D.M. Anderson, and T. Nemoto (Eds.), *Red tides: Biology, Environmental Science, and Toxicology*. Elsevier, New York.
28. **Bruno, D. W., G. Dear, and D. D. Seaton.** 1989. Mortality associated with phytoplankton blooms among farmed Atlantic salmon, *Salmo salar* L., in Scotland. *Aquaculture* 78: 217–222.
29. **Yagu Hansen, K.** 1989. [Toxic algae in Danish waters], pp. 21–29. in Report of the ICES Working Group on Harmful Effects of Algal Blooms on Mariculture and Marine Fisheries. *Int. Counc. Explor. Sea,* Doc C.M.1989/F:18, 80 pp.
30. **Imai, I., S. Itakura, and K. Itoh.** 1991. Life cycle strategies of the red tide causing flagellates *Chattonella* (Raphidophyceae) in the Seto Inland Sea. *Mar. Pollut. Bull.* 23: 165–170.
31. NOAA (National Oceanic and Atmospheric Administration). 1992. National statement: USA. Harmful algal blooms in the United States. Intergovernmental Oceanogr. Comm., Panel Meet. Harmful Algal Blooms, Paris, France, 5 pp.
32. **O'Mahony, J. H. T.** 1993. Phytoplankton species associated with imports of the Pacific oyster *Crassostrea gigas* from France to Ireland. *Int. Counc. Explor. Sea,* Doc. C.M.1992/F:26, 8 pp.; **Hallegraeff, G. M., D. A. Steffensen, and R. Wetherbee.** 1988. Three estuarine Australian dinoflagellates that can produce paralytic shellfish toxins. *J. Plankt. Res.* 10: 533–541; **Hallegraeff, G. M. and C. J. Bolch.** 1991. Transport of toxic dinoflagellate cysts via ships' ballast water. *Mar. Pollut. Bull.* 22: 27–30; **Hallegraeff, G. M. and C. J. Bolch.** 1992. Transport of diatom and dinoflagellate resting spores in ships'

ballast water: implications for plankton biogeography and aquaculture. *J. Plankt. Res.* 14: 1067–1084; **Bolch, C. J. and G. M. Hallegraeff.** 1990. Dinoflagellate cysts in recent marine sediments from Tasmania, Australia. *Bot. Mar.* 33: 173–192; **Kerr, S.** 1992. Ballast water — the next big shipping issue after oil spills? A summary of the Ballast Water Research Program in Australia.

33. **Burkholder, J. M., E. J. Noga, C. H. Hobbs, and H. B. Glasgow, Jr.** 1992. New "phantom" dinoflagellate is the causative agent of major estuarine fish kills. *Nature* 358: 407–410; **Burkholder, J. M., H. B. Glasgow, Jr., and K. A. Steidinger.** 1993. Unraveling environmental and trophic controls on stage transformation in the complex life cycle of an ichthyotoxic "ambush predator" dinoflagellate. Sixth Int. Conf. Toxic Mar. Phytoplankton, Nantes, France. Abstract. p. 43; Andersson, I. 1992. Aliens slip through the international safety net. *New Scientist* (July 3): 5; **Noga, E. J., S. A. Smith, J. M. Burkholder, C. Hobbs, and R. A. Bullis.** 1993. A new ichthyotoxic dinoflagellate: cause of acute mortality in aquarium fishes. *Vet. Rec.* 133: 48–49.
34. **Smith, S. A., E. J. Noga, and R. A. Bullis.** 1988. Mortality in *Tilapia aurea* due to a toxic dinoflagellate bloom, pp. 167–168. in *Proceedings of 3rd International Colloquium on Pathology in Marine Aquaculture*. European Association of Fish Pathologists, Aberdeen, Scotland.
35. **Shumway, S. E.** 1989a. A review of the effects of algal blooms on shellfish and aquaculture. *Int. Counc. Explor. Sea,* Doc. C.M.1989/E:25, 38 pp.
36. **Cosper, E. M., W. C. Dennison, E. J. Carpenter, V. M. Bricelj, J. G. Mitchell, S. H. Kuenstner, D. Colflesh, and M. Dewey.** 1987. Recurrent and persistent brown tide blooms perturb coastal marine ecosystem. *Estuaries* 10: 284–290.
37. **Glover, R. S., G. A. Robinson, and J. M. Colebrook.** 1972. Plankton in the North Atlantic — an example of the problems of analyzing variability in the environment, pp. 439–445. in Ruivo, M. (Ed.), *Marine Pollution and Sea Life*. Fish. News (Books) Ltd., London.
38. **Sheader, M. and F.-S. Chia.** 1970. Development, fecundity and brooding behavior of the amphipod, *Marinogammarus obtusatus*. *J. Mar. Biol. Assoc. U.K.* 50: 1079–1099.
39. **Powell, N. A., C. S. Sayce, and D. F. Tufts.** 1970. Hyperplasia in an estuarine bryozoan attributable to coal tar derivatives. *J. Fish. Res. Board Can.* 27: 2095–2096.
40. **Christmas, J. Y. and H. D. Howse.** 1970. The occurrence of lymphocystis in *Micropogon undulatus* and *Cynoscion arenarius* from Mississippi estuaries. *Gulf Res. Rep.* 3: 131–154.
41. **Linden, O.** 1976. Effects of oil on the reproduction of the amphipod *Gammarus oceanicus*. *Ambio* 5: 36–37.
42. **Woelke, C. E.** 1967. Measurement of water quality with the Pacific oyster embryo bioassay. Water Qual. Criteria, ASTM, STP416, pp. 112–120. *Am. Soc. Test. Materials,* Philadelphia, PA; **Woelke, C. E.** 1968. Application of shellfish bioassay results to the Puget Sound pulp mill pollution problem. *Northwest Sci.* 42: 125–133.
43. **Janssen, W. A.** 1970. Fish as vectors of human bacterial diseases. *Am. Fish. Soc.*, Spec. Publ. 5: 284–290; **Janssen, W. A. and C. D. Meyers.** 1968. Fish: serologic evidence of infection with human pathogens. *Science* 159: 547–548.

ADDENDUM:
INDICATORS OF HABITAT DEGRADATION

I have a brief addendum to this chapter on large-scale environmental events that can be considered possible indicators of widespread but not yet detectable changes in the coastal marine environment. Such events can often be spectacular and attention-getting, but *there are other less dramatic (and less speculative) indicators of habitat degradation* for which better evidence exists for a causal link with pollution. They all have to do, ultimately, with *communication*. To me, one of the most discouraging aspects of the ocean pollution problem is that we are receiving adequate warning about the amount of damage that is being done to the estuaries and the shallow edges of the sea and are ignoring the danger signs. The marine organisms whose immediate surroundings are being degraded are attempting to communicate their discomfort to us. The methods of communication, if we are perceptive enough to be aware of them and to interpret them, can provide us with an early warning system about increasing levels of environmental contamination. Some elements of the system are undoubtedly subtle and may well escape observation; others are relatively obvious.

Probably the best, and certainly one of the most apparent, indications of environmental degradation — whether by toxic or infective material, or chemical or thermal addition — is *mortality,* usually of a localized nature, in areas of heaviest contamination. Examples of this phenomenon are becoming increasingly abundant. As examples, repeated mortalities of fish and shellfish have occurred in Escambia Bay in northern Florida, a bay grossly polluted, principally by the poorly controlled effluents of several large chemical production plants, and in Puget Sound, Washington, heavily contaminated by industrial wastes. Beginning in 1967, summer fish kills occurred in Escambia Bay with increasing frequency and severity for a number of years, and in 1971 over 90% of the oyster population of that bay was destroyed within the space of a few days. Fish mortalities have been attributed to toxic chemicals, especially PCBs dumped in the bay and to low oxygen levels resulting from massive eutrophication. Mortalities of fish and shellfish have also been characteristic of other chemically polluted bays, such as Raritan Bay in New Jersey.

It is important to note, though, that mass mortalities due to natural causes are probably more abundant and more significant than those caused by human activities — man has merely added another group of stress factors.

The next best indication of environmental degradation takes the form of *drastic or subtle changes in the flora and fauna of an area,* either in terms of reduced abundance or in the disappearance or replacement of certain species. Generally, such changes can be summarized in three categories:

1. The decline and disappearance of fish species valuable as food or sport for man, and their replacement by rough species with lower value to man.

2. The development of a monotonous fauna consisting of fewer and more resistant species (such as certain annelid worms) that are able to tolerate low oxygen conditions.
3. Changes in the algal flora, often resulting in the appearance of blooms (frequently as red tides), and the predominance of blue-green and brown algae, the latter often occurring as a scum on inshore bottoms.

Occasionally, environmental changes may also result in population explosions of certain animal species that are normally inconspicuous parts of the fauna. An invasion of sea urchins in a sector of the Florida coast previously affected by a massive red tide outbreak in the summer of 1971 is a recent example. An earlier invasion by sea urchins occurred in kelp beds on the California coast.

An interesting report published over two decades ago[37] indicated that during the preceding two decades there had been a progressive decline in the abundance of many species and in the biomass of zooplankton in parts of the North Atlantic, together with a shortened season of biological activity. Among the many variables suggested as potential causes was the depressive effect of pesticides on phytoplankton photosynthesis. Such large-scale, long-term observations in the sea are all too rare, but they may indicate major derangements of man-made origin.

Another, and a more recently identified indicator of environmental contamination and degradation, is the *appearance of unusual or increased frequencies of abnormalities and diseases* in eggs, larvae, juveniles, and adults of estuarine and marine species. Documentation of this phenomenon is still very incomplete, but is adequate enough even at present to suggest that it will become a powerful tool in assessing the extent of damage to the marine environment caused by the effluvia of human civilization. Some of the varied forms include:

1. An apparent increase in observations of tumors and abnormal growths on fish taken from grossly polluted waters (information is available from California, New York, Massachusetts, Rhode Island, Washington, and Florida waters)
2. The appearance of fin and skin erosion — called “fin rot” — in fish from polluted waters (information is available from waters of the New York Bight, California, and Florida)
3. Erosion of the exoskeleton of Crustacea taken from polluted waters (information is available from New York Bight waters)
4. Increased frequency of fungus infections of eggs carried by Crustacea in areas of gross pollution[38]
5. Growth abnormalities in certain sessile invertebrates associated with chemical contaminants[39]
6. The appearance of lymphocystis (a viral disease of fish) in certain Gulf of Mexico estuaries with high pollution loads, and the absence of the disease in certain other, less-polluted areas.[40]

Several of these conditions (such as fin rot and lymphocystis) could result from increasing infection pressure by facultative pathogens, possibly combined with increasing environmental stress imposed by pollutants. Egg and larval abnormalities may also serve as sensitive indicators of environmental pollution. In the report just referred to on fungus infections in crustaceans,[38] studies in a bay on the coast of Britain disclosed that amphipods, *Marinogammarus obtusatus,* tended to be more abundant near a sewer effluent, and those nearest the effluent carried a much higher percentage of diseased eggs than those remote from the effluent (27% of a sample of 92 mature females vs. <1% of females from other parts of the bay). The authors suggested that microorganisms in the sewage may produce egg infections directly or that low salinities near effluents could kill the eggs or render them more susceptible to infection. As another example of the subtle effects on reproduction, a Swedish study on the amphipod *Gammarus oceanicus* showed that chronic low-level (0.3 to 0.4 ppm) exposure to crude oil resulted in a reduction of 20 to 50% in the numbers of offspring and in behavioral changes that reduce copulatory activity.[41]

Crustacean and molluscan larvae can serve as extremely sensitive indicators of environmental degradation. Use of larvae as bioassay organisms has a significant history on the West Coast in pulp mill pollution studies,[42] and is receiving increasing attention on the East Coast as well.

I have already suggested that bacteria of human origin may be facultatively pathogenic in stressed populations of marine animals, where they may produce effects unlike those produced (if any) in normal hosts. Organic loads from sewer effluents and sludge dumping promote bacterial growth, including that of heterotrophic marine or estuarine bacterial species, leading to tremendous infection pressure on fish and other animals by such facultative microorganisms. The suggestion has even been made, and some limited evidence exists,[43] that certain bacterial pathogens of humans are able to infect fish. Antibodies against such pathogens were demonstrated in fish from certain polluted waters in Chesapeake Bay, but not in those from relatively clean waters elsewhere in the bay. This work needs to be extended, but it does suggest that antibodies in fish may be used as sensitive indicators of pollution, whether or not the fish become grossly infected. There is also the likelihood that populations of bacteria such as the vibrios and pseudomonads, which may be pathogenic for humans who enter marine waters or eat the animals, may be enormously expanded by the availability of rich organic soups in outfall and sludge dumping areas.

In terms of impact on living marine resources, it seems reasonable to expect that the synergistic, cumulative effect of pollutants may well exceed the mere summation of individual effects. Thus, for example, chemical erosion of the mucus of a fish may expose it to invasion by facultative microorganisms; or modification of the physiology of a marine animal by high levels of heavy metals may lower its resistance to such facultative microorganisms.

It is likely that humans may serve as ultimate indicator organisms in assaying ocean pollution, since they are at the end of many ocean food chains, through fishing and mariculture activities. They may thus be exposed to concentrations of pollutants in their seafood (or other contacts with the marine environment) that may make them sick. A few such episodes, when brought to public attention, would do much to hasten measures to reduce ocean pollution.

To conclude this speculative discussion of large-scale and small-scale oceanic events that may be related to human interferences, I'd like to point out that the story of Chicken Little mentioned earlier has other lessons for us. You will recall from that timeless tale that the goose and the cow were swept along in the panic of the moment, possibly overreacting without analyzing the data set at hand, or without listening to the opinions of other presumed experts on atmospheric phenomena.

We should be a little brighter than they were in evaluating the severity of our own environmental crises — be they perceived or real. This we can accomplish by assessing the credibility of the people doing the analyses and giving us their interpretations, by trying to get some feeling for the robustness of the data on which conclusions are based, and by appraising the degree to which a cause-and-effect relationship has been demonstrated, since there are those who would argue with great vigor against efforts (such as this chapter) to assemble disparate events (such as ulcerations in fish and bleaching of corals) and to discern a trend or to implicate large-scale environmental influences. We still need to remember that natural systems are usually conservative; our responses to perceived crises should be the same.

It is important to recall also that modern predictions of future catastrophes are based on computer *simulations,* using whatever data are available. Scientists, using existing information, develop *scenarios* for the future based on chosen levels of critical variables. Unfortunately, with catastrophes, the experts are trying to model events that have never happened before, so the degree of variability is too large to ensure much validity in the predictions.

SECTION IV:
The Consequences of Inaction

Thus far in this consideration of coastal pollution and living resources, the focus has been on the data available to help assess effects of contaminants—effects on individuals as demonstrated by the appearance of diseases and abnormalities, and effects on populations as indicated by changes in reproductive potential and abundance. It seems logical at this point in the narrative to give some attention to what the future will be like in coastal waters, assuming that current trends, characterized by agonizingly slow responses to worsening environmental conditions, persist.

Factors critical to the continued use of coastal/estuarine waters for seafood production include demography—how many people choose to live in the coastal zone and what amenities they will demand there; economics—how much society is willing to spend on pollution abatement; and biology—essentially the degree of resilience of resource populations in the presence of habitat degradation.

This is a speculative exercise, but one for which some information of predictive value exists and has been presented in earlier sections of the book. Within the strictures of the data at hand, we can feel some minimal level of comfort in forecasting possible future trends based on a narrow choice of variables.

SECTION IV

11 Some Scenarios for the Future of Resource Populations in Polluted Zones

****

THE GREAT CONTAMINATED FISH SCARE IN JAPAN

Post–World War II industrial development in Japan has been a source of astonishment and envy for most of the rest of the world — but it has been achieved at a staggering environmental cost. Some elements of the cost became apparent on a local level during the Minamata Bay mercury poisoning incident of the 1950s and 1960s, but the true extent of damage to coastal waters was first impressed on the Japanese national consciousness in 1973, during what has been described as "the great fish panic." The episode was initiated by two reports that appeared almost simultaneously in the spring of that year. One, by a university medical research group, disclosed the discovery of new cases of mercury poisoning (Minamata Disease) in inhabitants of an area near the Ariaki Sea, 40 km north of Minamata Bay. The other disclosure, by the Japanese Fisheries Agency, was that PCBs were found to be above established human safety standards in more than 80% of fish sampled from six major coastal fishing areas.

Fear of the consequences of eating contaminated seafood, in a country whose animal protein source consists to a major extent of products from the sea, grew quickly and precipitated a violent and unusual response by the people. Within a few weeks, fish sales had plummeted by as much as 50%, and widespread mass demonstrations against polluting industries and inactive government agencies occurred. Dead fish were heaped in front of factory gates and government offices by angry fishermen, and thousands of placard-carrying consumers picketed the headquarters of the largest polluting industries. Media treatment was inflammatory, and inadequate solutions (government-sponsored surveys and industry promises to reduce contamination) were proposed to reassure an alarmed public.

But the event faded all too quickly — within two months. Newspaper coverage declined in favor of new crises of the moment (in this instance the hijacking of a Japanese commercial plane), and fish sales returned to normal levels. There were, however, a few persistent residues. One was heightened sensitivity of the average Japanese citizen to some of the real costs of industrial expansion — the consequences of the policy of economic growth at any price. Another was the realization that industrial pollution problems in coastal waters were national in scope, and required vigorous government intervention instead of the former laissez faire attitude of regulatory agencies.

From "Field Notes of a Pollution Watcher"
(C. J. Sindermann, 1981)

****

A favorite pastime among some groups of quantitative scientists — demographers, ecologists, and economists in particular — is to propose "scenarios" for future events, based on the selection of a stipulated set of conditions. Scenarios are descriptions of predicted consequences of actions or inactions taken or not taken within a specific system. This predictive exercise, in the right hands, can offer useful insights and is widely employed in population studies. It is of course subject to many limitations, including new ones outlined in recent discourse on "chaos" theory, in which minute variations can lead to major shifts in the outcome of any predicted interaction.

Scenarios proposed for future impacts of pollution on living marine resources must consider as a minimum the following variables:

- Future trends in levels of coastal/estuarine contamination in various geographic subdivisions of the world's oceans
- The likelihood, frequency, and possible locations of future acute contamination events, such as chemical spills
- The creation and deployment of new synthetic chemicals, which are often released without adequate testing on marine organisms
- The largely unknown sublethal effects of trace amounts of highly toxic synthetic chemicals already in the coastal environment
- The possible synergistic or antagonistic effects of complex mixtures of contaminants at different salinities and temperatures
- The true rate of degradation of organic synthetic chemicals under actual environmental conditions
- Short- and long-term changes in local and broad-scale hydrographic conditions

- Short- and long-term fluctuations in the abundance of marine populations, whether pollution-related or not
- The perception of risk by human populations, and the extent of their willingness to act aggressively in reducing coastal/estuarine contamination
- The kinds of market forces operating on seafood production to bring about change

This is my core list of variables to be considered in creating pollution-associated scenarios; other authors may have their own special lists, but they should include most of the above items, possibly augmented by some that I have not thought of.

Here is a small sample of some of the more obvious scenarios:

1. *A toxic event occurs, with significant human deaths and disabilities, as a consequence of ingesting contaminated seafood.*

One such catastrophe has already occurred: mercury poisoning of humans living near Minamata Bay in Japan, beginning in the early 1950s, as a consequence of eating contaminated seafood. Public indignation and legal measures resulted in industrial responses and stricter environmental regulations.[1]

With this scenario, any comparable future event will occur in a climate of already heightened public sensitivities to environmental abuses perpetrated by private industries. Public response will be massive, and severe governmental reactions will be immediate, leading to the assignment of full financial responsibility to the guilty industry, to a much-strengthened Environmental Protection Agency, and to implementation of rigorous federal guidelines for state environmental regulatory agencies. The episode will also begin a period of augmented funding for the nation's oceanic agency, with increased responsibility for monitoring contaminant levels in coastal/estuarine waters and in resource animals.

2. *Incidences of global pathological changes and mortalities in susceptible aquatic species will accelerate perceptibly and alarmingly as environmental breakpoints are reached and exceeded.*

The possibility was raised in Chapter 10 that at least some of the recent widespread pathologies and mortalities of certain marine species may be traced to increased stresses imposed by man's activities. Mentioned specifically were ulcerations in fish, bleaching of corals, mortalities of sea urchins, and mass deaths of marine mammals.

With this scenario, an expanding global human population and the absence of a strong worldwide commitment to reducing ocean pollution will lead to increasing stress on more susceptible species. This will result in additional reports of abnormalities and mass mortalities in more species, and even to the emergence and spread of new stress-related syndromes, such as widespread shell disease in crustaceans.[2]

Inability to establish clear cause-and-effect relationships between these events and increasing pollution levels will provide apologists with time to look for other scapegoats, and to delay the onset of genuine attempts to clean up coastal/estuarine waters.

The concept of "environmental breakpoints" is to some (but not all of us) the stuff of science fiction. The essence of the concept, to me, is that each species and each ecosystem has tolerance limits for particular injuries, chemical or otherwise. When enough of these limits are exceeded, the component parts (species) and eventually the entire system begins the process of collapse. This phenomenon may occur naturally, or it may possibly be precipitated by rising levels of toxic chemicals. The concept is obviously not very crisp, but I think a partial expression of it (as a consequence of *natural* stressors) occurred during the 1950s in the western North Atlantic — a time and place of widespread disease and death in marine species from unrelated pathogens. Sea herring died in great numbers in the Gulf of St. Lawrence;[3] wild and impounded lobsters died on the New England and Canadian Maritime coasts[4]; and mass mortalities of oysters began in Delaware and Chesapeake Bays.[5] It was as if a single environmental stressor — disease — had exceeded tolerance limits for several species simultaneously, and the coastal ecosystem itself was severely disturbed during that period.

3. *Worldwide occurrences of toxic algal blooms will increase in frequency and intensity, due to increased nutrient loading of coastal/estuarine waters and to global transport of nonindigenous toxic species — leading to permanent loss of some traditional fisheries and reduced availability of aquaculture sites.*

The early stages of this scenario are already in production; the only question remaining is the rate at which the process will accelerate. There has been, especially during the past three decades, a perceptible and global increase in the frequency and severity of toxic algal blooms, principally dinoflagellates.[6] Some have caused fish kills; some have damaged aquaculture production; and some have been toxic to humans. In the case of contaminated shellfish, toxins may remain at dangerous levels in the tissues for long periods — essentially removing the species from commercial utilization.

One significant observation has been that many of the recent blooms have been caused by algal species not previously known in the region affected but known elsewhere in the world. This has led to the suspicion (since substantiated) that algal species may be transported globally in ship's ballast water or as cysts attached to commercial products such as oysters and mussels that are a growing part of international trade.[7] Without stringent controls to prevent the introduction of such foreign species, more outbreaks will occur, more productive areas of estuaries and coastal waters will be seeded with populations of toxic algal species, and more species of shellfish (and even fish) will have to be withdrawn from commercial use because their flesh is toxic.

Another related observation with profound implications for the future is that modification of chemical balances in coastal/estuarine waters may result in large-scale and long-term changes in phytoplankton species, with the dominance of microalgae that are unsuitable as food for many herbivorous animals. Such changes could lead to the decline and even the disappearance of some of those herbivores in areas of former abundance. One dramatic example of just such effects was seen in the Long Island bay scallop mortalities of the late 1980s, when entire year-classes were obliterated by an extensive microalgal bloom.[8]

It is very easy to propose a scenario for a poisoned future in which marine aquaculture, touted for decades as a potential source of protein-rich food for an expanding human population, may by the year 2010 be seen to have fallen far short of its earlier promise. Major contributing factors will include increasing levels of coastal/estuarine pollution, which will reduce the growth and survival of captive fish and shellfish populations, and increasing occurrences of toxic algal blooms, enhanced by nutrients from human sources or derived from alien species imported accidentally from other parts of the world ocean.

It is possible to envision an even more dismal near-term situation — within two or three decades — in which farming the edge of the seas for salmon, shrimp, scallops, mussels, and other species will have to be abandoned in some geographic areas because of recurrent — even annual — blooms of toxic algae, including widespread distribution of the recently recognized "phantom algae" described in Chapter 10 that are so lethal to fish.

More disturbing from a public health perspective may be episodes of human illness and deaths in coastal communities, especially among the elderly and those with respiratory problems, resulting from breathing airborne molecules of toxins. Alarmists, including some members of the scientific community, even foresee cessation by the middle of the twenty-first century of all human utilization of natural and aquaculture production of fish and shellfish in some coastal/estuarine waters because of high tissue levels of toxicants of natural or industrial origin. High seas stocks of tuna and other large predators would not be exempt either, without elaborate testing for contaminants before marketing.

4. *A major research and monitoring effort by a revitalized national oceanic agency concludes with the clear verification of significant negative effects of coastal pollution on the abundance of living resources.*

The availability of palatable uncontaminated seafood is of concern to most consumers. Clear evidence that resource species are being damaged severely by coastal pollution will galvanize public support for stronger governmental actions to reduce contamination of productive waters. The often-used argument that we in the U.S. can import seafood from other parts of the world will no longer be valid, since those areas are becoming increasingly crowded and increasingly polluted as well.

Substantial legally viable new evidence of damage to resources will be used effectively by a much-strengthened Environmental Protection Agency to deal more harshly with polluters, most of whom are known. Stronger measures will be used against coastal municipalities such as New York and Boston that have moved reluctantly in pollution abatement matters. Actions to prohibit forever the abomination of ocean dumping will also be aided by the new evidence of damage to resources.

5. *Levels of toxic chemicals in coastal/estuarine waters increase gradually, in synchrony with the expansion of human populations in adjacent coastal zones. As contaminant concentrations in fish and shellfish tissues approach action levels, regulatory agencies impose bans or other increasingly severe restrictions on fisheries.*

Coastal pollution can affect the availability of seafood in human diets in two important ways: (a) it may increase mortality or decrease fecundity of fish and shellfish, resulting in smaller resource populations; and (b) even if fish and shellfish abundance is not affected and no mortalities occur, contaminants in tissues may render seafood inedible or unsafe for human consumers. It is likely that both of these actions can occur simultaneously in localized areas of severe contamination, but the regulatory response will focus on threats to public health rather than to effects on the well-being of the living resources.

Elements of this scenario are already part of the regulatory reaction to the existence of "action levels" of certain contaminants in fish. PCB contamination of fish in the Hudson River and its estuary in the 1970s and 1980s led to a ban on fishing for and possession of species such as striped bass, eels, and others.[9] Earlier, the presence of high tissue levels of mercury in large predatory fish such as tuna and swordfish resulted in similar bans.[10]

6. *No major ocean pollution crises develop. Present discharge practices continue to be used by expanding coastal human populations, and amounts of contaminants in coastal waters and in seafood are worrisome, as reports of high levels persist, but life goes on.*

We may not like to admit it, but this scenario has to be voted the most likely to occur. We already have too many environmental crises to occupy our attention — global warming, acid rain, deforestation, ozone depletion — each with proponents for remedial action running hard to obtain attention and support. One more crisis, and admittedly one without a sharp focus for action, is apt to be back in the pack with the other contenders.

Minor flaps involving localized seafood contamination will undoubtedly continue, and regulatory agencies will briefly tighten surveillance measures. Some people will get sick from eating raw, microbially contaminated seafood, and most people will be regularly or intermittently ingesting low levels of toxic chemicals with their seafood, but the effects will not be dramatic enough or life-threatening enough to engender any strong response.

At some future point in this scenario, though, seafood consumers will realize that they are increasingly at risk from toxic contaminants in food, as did the Japanese during the great fish scare of 1973. Several parallel events should then occur: (a) decreasing market interest in suspect seafood products, (b) demands by seafood consumers for more stringent inspection procedures and sanitary standards, and (c) gradual withdrawal of fish and shellfish production from contaminated coastal/estuarine waters into land-based aquaculture facilities in which clean water can be guaranteed.

Although more and more people are better informed about the polluted state of their environment, many have yet to be convinced that the situation warrants concerted action. Optimists among us cling to the idea that it is possible, through education, to bring the state of public thinking to a point where progress can be made. Others feel that crises are the only effective motivators — and may remain so.

So here then are six scenarios — (1) a major toxic event, Minamata-level; (2) possible global impacts of exceeding environmental breakpoints; (3) pollution-associated toxic algal blooms; (4) consequences of demonstrating pollution impacts on resource abundance; (5) regulatory responses to contamination of seafood; and (6) the consequences of "laissez-faire" policies concerning coastal pollution. Once the flavor of scenario creation has been sampled, the possibilities for additional scenes blossom endlessly. The six just considered are probably the most obvious possibilities, but there are many others that could be explored (but not here).

Scenarios are the proper playthings of quantitative scientists, and are based on manipulations that begin with the development and validation of *models,* which are numerical representations of the real world. By changing the inputs, boundaries, and even the structure of a model, scientists can produce *simulations,* which are computer experiments based on different combinations of conditions. Any one combination of variables may then be selected as the basis for a *scenario,* which can be defined as "a specific combination of inputs and system properties and the resulting outputs".[11] The "outputs" can portray conditions and events in the future (based on the stipulated set of conditions) with a view, in the case of coastal pollution, toward prevention or mitigation. The attractiveness of scenarios as projections of future events is reduced significantly, in my estimation, by the intrusion of "chaos" into simulation models, by the possibility of a major unpredicted event (such as the Minamata incident), and by the limited availability of much supporting quantitative data on pollution effects.

REFERENCES

1. **Harada, M.** 1972. *Minamata Disease*. [In Japanese.] Iwanami Shoten, Tokyo, 274 pp.; **Huddle, N. and M. Reich.** 1975. *Island of Dreams: Environmental Crisis in Japan*. Autumn Press, New York. 225 pp.; **Smith, W. E. and A. M. Smith.** 1975. *Minamata*. Holt, Rinehart and Winston, New York. 220 pp.
2. **Sindermann, C. J.** 1989. The shell disease syndrome in marine crustaceans. U.S. Department of Commerce, Natl. Mar. Fish. Serv., Woods Hole, MA. NOAA Tech. Memo. NMFS-F/NEC-64, 43 pp.
3. **Leim, A. H.** 1956. Herring mortalities in the Gulf of St. Lawrence, 1955. *Misc. Spec. Publ., Fish. Res. Board Can.* 607: 1–9; **Tibbo, S. N. and T. R. Graham.** 1963. Biological changes in herring stocks following an epizootic. *J. Fish. Res. Board Can.* 20: 435–449.
4. **Goggins, P. L. and J. W. Hurst, Jr.** 1960. Progress report on lobster gaffkyaremia (red tail). *Maine Dept. Sea Shore Fish. Bull.*, 9 pp. (unpubl. mimeo.).
5. **Andrews, J. D.** 1979. Oyster diseases in Chesapeake Bay. *Mar. Fish. Rev.* 41: 45–53; **Haskin, H. H. and S. E. Ford.** 1983. Quantitative effects of MSX disease *(Haplosporidium nelsoni)* on production of the New Jersey oyster beds in Delaware Bay, U.S.A. *Int. Counc. Explor. Sea,* Doc. C.M.1983/Gen:7, 21 pp.
6. **Hallegraeff, G. M.** 1993. A review of harmful algal blooms and their apparent global increase. *Phycologia* 32: 79–99.
7. **Hallegraeff, G. M. and C. J. Bolch.** 1991. Transport of toxic dinoflagellate cysts via ships' ballast water. *Mar. Pollut. Bull.* 22: 27–30; **Hallegraeff, G. M., C. J. Bolch, J. Bryan, and B. Koerbin.** 1990. Microalgal spores in ships' ballast water: a danger to aquaculture, pp. 475–480. in Granéli, E., B. Sundström, L. Edler, and D.M. Anderson (Eds.), *Toxic Marine Phytoplankton. Proc. Fourth Int. Conf. Toxic Mar. Phytoplankton, Lund, Sweden*. Elsevier, New York.
8. **Cosper, E. M., W. C. Dennison, E. J. Carpenter, V. M. Bricelji, J. G. Mitchell, S. H. Kuenstner, D. Colflesh, and M. Dewey.** 1987. Recurrent and persistent brown tide blooms perturb coastal marine ecosystem. *Estuaries* 10: 284–290.
9. **Nadeau, R. J. and R. A. Davis.** 1976. Polychlorinated biphenyls in the Hudson River (Hudson Falls–Fort Edward, New York State). *Bull. Environ. Contam. Toxicol.* 16: 436–444; Anonymous. 1976. PCB's contaminate freshwater fish. *Va. Mar. Times* (summer 1976), p. 2.
10. **Taub, H. J.** 1973. Tuna protects against mercury. *Prevention* (May 1973), pp. 29–38; **Officer, C. B. and J. H. Ryther.** 1981. Swordfish and mercury: a case history. *Oceanus* 24(1): 34–43.
11. **Karplus, W. J.** 1992. *The Heavens are Falling: The Scientific Prediction of Catastrophes in Our Time*. Plenum Press, New York. 320 pp.

SECTION V:
Proposals for Change

The evils of coastal/estuarine pollution, as outlined in earlier chapters of this book, include (but are certainly not limited to) increased environmental stress on resource species, with concomitant increases in prevalences of certain pollution-associated diseases of fish and shellfish; greater risk of human infections, especially from ingesting raw, contaminated shellfish; and reduced reproductive potential in some species from heavily contaminated areas, with possible localized effects on population abundance. This litany seems adequate to support a perception that we as the culprits (and ultimate victims) should be doing something, or more than something, to address the problem of ocean pollution, especially when it impinges on our shores and contaminates our seafood.

The period of earth's history when the human species has been abundant enough to make any impact on the oceans is vanishingly brief, and the period when members of the species became aware of that impact is but a tiny fraction of that brief period—extending back about a century for scientific observations of impacts of fishing on local stocks, and only about three decades for the detection of measurable industrial pollution in localized parts of the coastline.

Proposals have been made throughout this period of growing awareness for actions to reduce damage to marine ecosystems and particularly to living resources. Fishing regulations were developed for some overfished stocks, and international commissions were organized to reduce overexploitation of high-seas stocks. Coastal pollution had fewer champions, except where abuses were obvious and the guilty parties could be pursued into the courts. International conferences were held to discuss proposals for pollution limitation, with little concrete action. Regional conferences and subsequent commissions—such as those for the Baltic Sea and the North Sea—produced more specific results, once the consequences of inaction were recognized.

The reality is, though, that despite good intentions, conferences, commissions, and regulations, the human impacts of overfishing and pollution on coastal resource populations and ecosystems are still very much a problem. Proposals for pollution abatement have met with great resistance from polluters, and some nonpoint sources (such as air pollution) have been remarkably unresponsive to control measures. So, despite a long record of unfulfilled proposals by others, I have developed a short list of my own (although the individual items are by no means original). This list is explored in Chapter 12.

12 Reducing Contaminant Levels in Coastal/Estuarine Waters

****

EARTH DAY IN FLORIDA

It was a sparkling late April morning in 1970, in Bayfront Park on the Miami city waterfront. Hippies and homeless had been sequestered by the police in a far corner of the park and a platform had been built to elevate inspirational speakers above the crowd in celebration of the first-ever "Earth Day." The heads of all the local agencies and institutions with any environmental involvement were present — the Marine Institute of the University of Miami, the Tropical Atlantic Biological Laboratory, the Atlantic Oceanographic and Meteorological Laboratory, the National Hurricane Research Center, the Agricultural Plant Introduction Station, the International Oceanographic Foundation, and many others. Each institutional representative had been allotted precisely five minutes to say something meaningful (and newsworthy) about the environment to an assembled audience of several hundred.

The speeches reflected recently discovered environmental sensitivities — even mine, although I admit with some chagrin that as director of the Tropical Atlantic Biological Laboratory my attention was then focused more on American tuna fisheries off the West African coast rather than on coastal pollution problems at home. Two widely publicized oil spills the year before, at Santa Barbara, California, and West Falmouth, Massachusetts, had exacerbated concerns about human actions harmful to world ecosystems, and every speaker emphasized the distressed state of the environment and the need for remedial measures. I didn't learn until much later that one of my sons and some of his friends from Coral Gables High were, quite by accident, standing in the crowd. He didn't expect to find me on the platform as part of the program, and I hadn't appreciated his early involvement in ecological concerns. I guess it was one of those mutual growth experiences for both of us that are too rare during teenage years.

Because of the coastal venue of the ceremony, emphasis was logically on the state of the marine environment, particularly on the polluted condition of Biscayne Bay, conveniently located only a few steps from the rear of the speakers stand. At any rate, I think that most of us went away from that gathering with some kind of vague resolve to pay more attention to environmental happenings around us, and to become parts of solutions rather than problems, insofar as human habitats were concerned. Those bright commitments dimmed with time of course, but a residue of heightened awareness has persisted — so that senseless degradation of coastal waters is no longer as easy to justify in 1980 as it was a decade earlier. Biscayne Bay still can't be described as pristine, and developers still occasionally win legal battles that allow them to destroy coastal wetlands, but regulatory agencies are now more responsive to public pressures, and media reactions to environmental abuses are prompt and incisive.

From "Field Notes of a Pollution Watcher"
(C. J. Sindermann, 1980)

****

Contaminant levels in estuarine/coastal waters have been measured by numerous surveys and assessments conducted during the past two decades, and monitoring programs such as "Mussel Watch" and NOAA's "Status and Trends" program have broadened the geographic and temporal horizons of our information. As the data accumulate, it is becoming increasingly clear that contamination of coastal waters is widespread, but that localized areas of heaviest impacts on living resources exist near centers of human population and intensive industrial development. These are the areas that require concerted efforts to reduce contaminant loading, but these are also the areas where pollution problems seem most resistant to rapid solution.

Every author who has examined and written about environmental degradation resulting from human activities has, usually near the end of his or her treatise, proposed a suite of remedies — a listing of steps to alleviate perceived damage. I have such a listing. Its elements are certainly not original, and they all depend for implementation on a *public perception of the reality of danger* and the *public will to correct abuses*. They depend too on accepting the inevitability of stringent regulatory measures that are rigidly enforced and that may be costly.

Some critical steps are these:

1. Prohibit — absolutely — all ocean dumping of toxic wastes and sludges.
2. Require that all effluents from ocean outfalls be of drinking water quality.

3. Resist further development of estuarine and shoreline areas unless such development is ecologically sustainable.
4. Accelerate river and watershed cleanup, and encourage agricultural and other industrial practices that are critical to the survival of productive estuaries.
5. Designate and protect coastal/estuarine sanctuary areas for aquaculture production.
6. Encourage public demands for an aggressive yet responsive national environmental protection agency, equipped with sufficient legislation to emancipate it from the whims of any reactionary politician and with adequate resources for enforcement of regulations.
7. Encourage the United States, through political pressures and congressional action, to assume a dominant, aggressive international leadership role in reducing ocean pollution.

At first glance, some of the items in this short list seem idealistic and almost impossible to achieve, especially when so many other environmental crises jostle for our attention and action. This may be true, but the alternative — progressive deterioration of an important component of the human environment — is totally unacceptable to most of us. The country's estuaries and coastlines deserve a better fate than as collection sites for noxious material generated by an uncaring industrial society.

Before exploring each of these proposals for environmental action in some detail, the critical role of public perception and public will should be reemphasized, since without these ingredients little can be accomplished. The truth of this observation is evident in the actions (or inactions) of America's principal environmental protector — the U.S. Environmental Protection Agency — during the four decades of its existence. Under some political administrations, particularly that of Ronald Reagan, the agency was virtually inactive (many would say retrogressive). The pace of developing guidelines and enforcing existing ones obviously depended then and now on public pressures as exerted through news media and legislatures. If the public is not concerned and vocal, then the agency and the politicians are not concerned either, and environmental abuses persist and intensify. If, on the other hand, the EPA looks over its collective shoulder and perceives a substantial mass of aroused citizens behind it, then corrective activities seem to be enhanced. Transient, ultraconservative political administrations, such as that of former President Reagan, can briefly interfere with progress in environmental protection, but the long-term trend is encouraging, although to some it seems exasperatingly slow. (It is worrisome to note, though, that now, in early 1995, a Republican-dominated Congress is bringing forth legislation to "reduce the regulatory burden on industry and agriculture" — by gutting many of the country's most important environmental laws.)

With a concerned public and a revitalized EPA as active participants, we can now turn to the list of proposed measures to reduce pollution levels in coastal/estuarine waters. The elements of the proposals should, of course, form an integrated whole, even though each can be a force for improvement in its own right.

1. PROHIBIT OCEAN DUMPING

During the decade of the 1970s, I was director of the Marine Biological Laboratory at Sandy Hook, New Jersey, within sight of the dim, smog-enshrouded skyline of New York City across the harbor. That decade was an exciting one for many of us who were involved in environmental studies. Earth Day — the original one — had just happened, and sensitivities to human impacts on ecosystems had been at least temporarily heightened. One of the principal research programs of the Sandy Hook Laboratory was to examine the effects of coastal pollution, including that from ocean dumping, on fish and shellfish populations and their habitats. Our attention was focused on a sludge dumping site 11 miles east of Sandy Hook, a revolting place that, like war, is almost impossible to describe. Sludge barges from New York and New Jersey sewage treatment facilities made daily trips to this foul piece of ocean and added their loads to an area known locally as "the dead sea." The location was easy to find because of the smell and the residuum of "floatables" from the sludge, and was even visible from satellites as an area of U-shaped oil slicks and brown water (Figure 38). The sea bottom there was heavily impacted and devoid of most of the usual species of bottom-dwelling animals. Fish from that zone displayed eroded fins and ulcerations, and crustaceans often had blackened and eroded spots on their shells.

FIGURE 38. High-altitude photograph of the former sludge dump site 11 mi east of New York City. The U-shaped slicks were formed as sludge barges dumped half their load on the outward leg and the other half of their load on the return leg of the trip.

Ocean disposal of human fecal wastes has been practiced since the invention of the flush toilet in the nineteenth century by Sir Thomas Crapper, and ocean dumping of concentrated sludge has been a favorite choice of sanitary engineers since the development of sewage treatment plants. The prevailing premise had always been that the ocean was so vast and its assimilative capacity so great that no insult by humans could make an impact. But research at Sandy Hook and elsewhere began to demonstrate that this premise was incorrect and that it was possible to abuse coastal/estuarine waters near centers of population to such an extent that their productive systems were degraded. Evidence of damage to resources and habitats in the New York Bight and at other coastal dump sites forced a reluctant EPA and resistant municipal agencies (especially New York City) to consider, in the mid-1980s, cessation of ocean dumping.

A temporary alternative was found, though, in merely moving the sludge dumping area out 106 miles from New York, to the edge of the continental shelf, where its effects would be less visible. Persistent NOAA and academic scientists began finding evidence of effects on resource animals even at that remote location, and only then were definitive steps taken by EPA in 1991 to halt all sludge dumping at sea — still with last-ditch delaying tactics by New York City and New Jersey sewerage authorities. The 106-mile site was finally closed to further sludge dumping in 1992.

The reality is that ocean dumping — whether it be of sewage sludge, contaminated dredge spoil, industrial wastes, or refuse from construction and city streets — must cease. It is a degrading process, inimical to uses of coastal/estuarine waters for seafood production, despite lingering ideas that such waters have great assimilative abilities, and despite the potential for the use of sludges to enrich relatively unproductive areas (a potential that has not been realized, or even tested adequately, in the sea). Ocean dumping represents the lowest possible use of the sea — as the recipient of the offal of human populations. We can do better!

2. REQUIRE THAT ALL EFFLUENTS FROM OCEAN OUTFALLS BE OF DRINKING WATER QUALITY

I have just properly characterized ocean sludge dump sites as "foul revolting places," but much of the same language could be used to describe so-called "ocean outfalls" of municipal sewerage systems — giant pipes that extrude inadequately treated or untreated human wastes into harbors (Boston, New York) or shallow coastal waters (New Jersey, Florida, California).

The horror stories associated with ocean outfalls are numerous; a few examples will give a little flavor here:

1. California has some of the biggest outfall pipes in the world. An early study, 40 years ago, of the effects of sewage effluents from one Los Angeles outfall[1] disclosed severe effects on fish in the immediate area. Many had eroded fins and ulcerations, and their flesh was flaccid and watery. Some had skeletal defects and missing eyes; others had large external tumors.
2. Large ocean outfalls are also abundant along the Florida coast. One, now defunct, off 79th street in Miami Beach, created a giant circular brown stain in the otherwise blue-green coastal waters. The brown area was dubbed "the rose bowl," and its location was very apparent from the air — to tourists and others escaping from frozen northern cities in winter. Its foul smell was distinctive from surface craft, even though the party boat fishing in the immediate area was reported to be good, and maybe some of the fishermen actually ate what they caught in those degraded waters.
3. The shoreline of New Jersey — from Sandy Hook south to Atlantic City — is pierced at irregular intervals by ocean outfalls extending for variable but always inadequate distances from shore. Public health officers of most coastal municipalities in this much-abused, overpopulated part of the nation have to be really inventive in the summertime to prevent closure of their beaches because of pollution (largely from these outfalls). One favorite device of some communities is to store great quantities of sludge during the summer and then flush it through the outfall pipes during the cooler period when fewer people are on the beaches to witness the "floatables" or to contract various illnesses from contact with viral and bacterial contaminants.

Reading about these and other degrading practices in our coastal waters, it is difficult not to ask "why?". Why in a wealthy, technologically advanced country such as ours do we condone such primitive and needless abuses? Why do we not improve sewage treatment facilities in the coastal zone (as has been done elsewhere in the nation) so that we can remove forever this crutch of ocean disposal? The answer of course is cost. Advanced sewage treatment facilities are expensive — even when balanced against barging sludge out to the edge of the continental shelf. Municipalities perpetually plead poverty, and expect federal funds before actions are taken, so any definitive steps are deferred in favor of the status quo (which is not really status quo when confronted with increasing human pressures on the coastal zone).

3. RESIST FURTHER SHORELINE DEVELOPMENT UNLESS IT IS "ECOLOGICALLY SUSTAINABLE"

Shorefront land and coastal wetlands continue to be modified for human purposes — mostly housing — in many states despite sporadic attempts by local jurisdictions to slow the pace. Part of a formula for resisting further poorly planned exploitation of coastal lands and marshes is universal endorsement of and adherence to the emerging concept of *ecologically sustainable development*.

"Sustainable development has been defined as *development that meets the needs of the present without compromising the ability of future generations to meet their own needs*."[2] The *ecological* perspective on sustainable development focuses on natural biological processes and the continued productivity and functioning of ecosystems. Long-term ecological sustainability also requires protection of genetic resources and conservation of biological diversity.[3]

One major sustainability concern (in a list of some half dozen) is that "Humanity's wastes are accumulating to such an extent as to severely compromise future use of the biosphere".[4] Ocean pollution is one example of the reality of that concern. It is part of an unprecedented dissatisfaction with the abusive behavior of the human species, insofar as environmental management practices are involved. Part of the concern is also that unplanned development may be the greatest single factor (except possibly for overexploitation) affecting fisheries of the U.S. Estuarine dependency of young fish occurs in 95% of species in Atlantic coastal states.

Barriers to sustainable fish production include freshwater diversions, eutrophication, chemical contamination, and physical modification of habitats. But the effects of these barriers at the population level have not been determined with any degree of precision. Until they are, impacts on fish abundance will remain matters of conjecture.

4. ACCELERATE RIVER AND WATERSHED CLEANUP

It is becoming clear that limiting ocean pollution will require control of events that take place on land — even in locations remote from coastal areas. Pollution of watersheds has visible effects on rivers and eventually on their estuaries. With these obvious facts in mind, environmental cleanup programs such as the EPA-sponsored Chesapeake Bay Program, with the objective of reducing pollutant inputs, have adopted a holistic approach, recognizing the bays and estuaries as part of a much larger interacting ecosystem. Events in areas far removed from the coast — in remote watershed sites — can influence the well-being of living organisms in the lower reaches of the system. So, for example, fertilizers, pesticides, mine washings, and industrial chemical wastes that enter streams and rivers will eventually have an impact in the estuary. Furthermore, airborne contaminants from agricultural applications or from polluting industries, add significantly to pollution problems in the lower reaches of the system. To add to the collective insult, most large coastal cities are located on bays or estuaries, where they make a major contribution to aquatic environmental degradation.

Successful mitigation programs, such as the Chesapeake Bay Program, have already demonstrated that regional or whole watershed programs are the most efficient. Extremely local problems (such as Baltimore Harbor) need to be addressed as well, but they should be approached from a broader perspective and as components of broader programs. A critical mass of competencies to address environmental issues is more apt to be available regionally than locally. But — and this is a large but — in the pursuit of regional cooperation, it is important not to lose sight of the fact that the region is part of a still-greater whole. A geographically well-defined coastal region such as Chesapeake Bay still experiences problems whose boundaries extend beyond the region and therefore are not easily manageable by the coastal states alone. Examples would be air pollution, or the management of large game fish and other highly migratory species. For these we clearly need programs that are national in scope.

To meet the needs of pollution abatement programs, whether they be local, regional, or national, there is the never-ending need for data — especially that derived from environmental and resource monitoring efforts. Fortunately, in many sectors of the U.S. coastline, there are at present *extensive* as well as *intensive* programs monitoring human-mediated and natural environmental changes. Of large geographic scale, for example, is the NOAA Status and Trends Program, a nationwide monitoring effort to measure the biological significance of contaminants. Participants in the program examine sediments, shellfish, and bottom-dwelling fish for toxic chemicals. This is a multilaboratory program, designed to be long term. It uses an array of indicators; it continues to explore new approaches; and it is flexible enough to change when advisable. At the other end of the spectrum are the intensive studies of localized areas, such as a multidisciplinary study of two South Carolina estuaries — Murrills and North Inlets. Nutrients, sediment chemistry, hydrography, oyster contaminant levels, and microbiology are all parts of the research effort.

We need more of these extensive and intensive examinations of environmental status and changes in status, since the quality of management decisions about pollution reduction can be improved by data from them. Programs need to be long term and in multiple systems. They need to focus on the severely impacted areas, but they must also transcend such a narrow focus for greatest utility.

5. DESIGNATE AND PROTECT COASTAL/ESTUARINE SANCTUARY AREAS FOR AQUACULTURE PRODUCTION

Marine aquaculture, particularly of shrimp, oysters, scallops, mussels, and salmon, is expanding rapidly in many parts of the world and now contributes over 15% of global seafood production. It is amazing to me that such expansion has been possible despite the absence of effective national or international control of coastal/estuarine pollution. Here and there, though, there are predictive cases of negative impacts of industrial pollution on aquaculture. Japan, with major dependence on marine fish and shellfish as protein sources, has produced glaring examples of the incompatibility of aquaculture and unregulated, polluting industries. Expanding chemically contaminated zones now surround its major coastal cities such as Tokyo, Osaka, and Hiroshima, and most of its smaller coastal cities as well.[5] Pollution of a different kind — human fecal contamination — has affected shellfish production in Italy and Spain by causing sporadic disease outbreaks among seafood consumers. Nations such as Japan, where coastal aquaculture has been severely impacted by existing pollution levels, have slowly and reluctantly responded with increasingly severe restrictive legislation and regulations.

Marine aquaculture production in the U.S. has developed at a slower pace than in many other countries, but, even so, problems with coastal/estuarine pollution have forced the closing of shellfish producing areas and abandonment of some aquaculture sites because of agricultural or other industrial pollution. Until recently, enhanced environmental sensitivities resulted in closer examination of long-term damage from chemical pollution — priority was rarely given in this country to the absolute requirement for clean water for aquaculture ventures.

We now need a much more aggressive national stance toward marine aquaculture development and toward protection of the coastal/estuarine waters where cultured animals will be grown. Current public attitudes favor both activities; they are closely intermeshed; in fact, aquaculture development can be a major factor in reinforcing the need for pollution control.

6. ENCOURAGE PUBLIC DEMANDS FOR ENVIRONMENTAL PROTECTION

A small number of industry executives and industry-supported politicians, pursuing narrow financial and political goals, have for many decades participated in deliberate polluting actions that have helped to deprive the rest of us of clean beaches, safe seafood, and marine resources for the future. In some instances, their destructive practices continued long after evidence of environmental damage became apparent, while scientifically trained apologists in the employ of the polluting industries obscured or denied harmful effects. Vigorous regulatory steps or costly legal suits are usually required to bring about change, and even these measures are slow to correct abuses. Informed, adequately financed, and aggressive public interventions, leading to rigidly enforced regulations with severe penalties for polluting, are required.

Reducing coastal/estuarine pollution is a major and difficult responsibility of government regulatory agencies at all levels — federal, state, and local — and can be best accomplished with the support and participation of an informed public. Although much has already been accomplished during the past two decades, there is a continuing need to further develop *constituencies* — private citizens and environmental defense organizations — committed to opposing abuses of coastal/estuarine waters and habitats of resource species. Part of the commitment is of course financial. We are all "environmental consumers," so we must be ready to pay (through tax increases) to solve some of the environmental problems that we and our polluting industries have helped to create. Awareness of the existence of such problems must be increased, with the assumption that the more informed the taxpayers are, the more they will value their natural resources, the more they should be willing to pay in taxes and costs of goods to protect those resources, and the greater their indignation will be against those who persist in despoiling their oceans.

A critical component of any public response must be *support of strict enforcement of existing environmental regulations*. A complex mix of government agencies is involved in enforcing pollution-control legislation and regulations. Coastal and estuarine waters are principally within state jurisdictions, so much of the enforcement responsibility is the purview of state departments of environmental protection or interstate commissions given authority by the member states over specific bodies of water. County and municipal governments may in some states have jurisdiction in local aquatic environmental matters. National agencies such as the Environmental Protection Agency, the Food and Drug Administration, and the Coast Guard also have enforcement responsibilities in specific areas mandated by federal legislation. The existence of all these layers of regulatory responsibilities does not, however, ensure that laws and regulations will be enforced, unless public concern is evident and continuing financial support is available. The principal reason given for inadequate enforcement is always lack of adequate resources — funding and people — to carry out the necessary control functions. This excuse for inaction should be eliminated.

7. ENCOURAGE THE U.S., THROUGH POLITICAL PRESSURES AND CONGRESSIONAL ACTION, TO ASSUME AN AGGRESSIVE INTERNATIONAL LEADERSHIP ROLE IN REDUCING OCEAN POLLUTION

Some of the earlier proposals in my short list of methods to reduce ocean pollution — such as expecting effluent pipeline discharges to be drinkable (or at least of swimmable quality) — risk being dismissed as idealistic or improbable (or even facetious). Such is not the case with the final recommendation that this country should assume a strong role in promoting pollution reduction in its home waters as well as globally. The list of conventions and treaties to which the U.S. is signatory is extensive — beginning 40 years ago with the 1954 International Convention for the Prevention of Pollution of the Sea by Oil (London), and including the 1975 International Convention on the Prevention of Marine Pollution by Dumping of Wastes and Other Matter (the so-called London Ocean Dumping Convention). This important convention prohibited ocean dumping of toxic substances such as organohalogen compounds, mercury and mercury compounds, cadmium and cadmium compounds, as well as plastics and other persistent synthetic materials. The landmark 1972 U.N. Stockholm Conference on the Human Environment concluded with a strong declaration against ocean pollution; this position was reaffirmed in a lukewarm way two decades later at the recent (1992) U.N. Rio de Janeiro Conference.

A global perspective on future developments, including pollution control, by the World Commission on Environment and Development (1987),[2] included the needs for:

- A *production system* that respects the obligation to preserve the ecological base for future development
- A *technological system* that can search continuously for new solutions to problems such as coastal pollution
- An *international system* that fosters ecologically sustainable patterns of trade and finance

The U.S. is fortunate to have political, social, and economic systems required to address these needs, once we have concluded that ocean pollution reduction is a desirable goal, nationally and internationally.

To conclude this discussion of proposals for change, I want to return to the listing of global environmental concerns mentioned earlier in this chapter.[4] One overriding concern is that "Resources are being used or degraded at such a rate as to significantly compromise their availability to future human generations." We think of living marine resources as renewable, unlike fossil fuels or other mineral deposits. If, however, we degrade habitats sufficiently, reducing reproduction or survival of resource populations, then we assume the power to diminish or eliminate renewability of those resources. We should not allow that to happen.

This concern about the future has been expressed in larger global circles in a slightly different way, pointing out that "*our current organization of society and modes of production and consumption are not sustainable,* that we as a species constitute a destructive ecological force, and that we need to undergo a transformation in thinking, planning, and acting, so we can meet the needs of the present without compromising the ability of future generations to meet their own needs" (a very rough quote from the World Commission on Environment and Development, 1987).[2] Surely, the recent history of our mistreatment of coastal/estuarine waters attests to a need for change. It is apparent — at least to me — that major long-term commitments are necessary. We need specific remedial actions in support of decades of rhetoric about pollution abatement. René Dubos, in his 1981 book *Celebrations of Life*[6] has pointed the way:

> ...it is possible for human beings to modify the surface of the earth in such a way as to create environments that are ecologically viable, esthetically pleasurable and economically profitable....

The coastal oceans are in desperate need of the application of such a philosophy.

REFERENCES

1. **Young, P. H.** 1964. Some effects of sewer effluent on marine life. *Calif. Fish Game* 50: 33–41.
2. World Commission on Environment and Development. 1987. *Our Common Future*. Oxford University Press, Oxford.
3. **Brown, B., M. E. Hanson, D. M. Liverman, and R. W. Meredith, Jr.** 1987. Global sustainability: toward definition. *Environ. Manage*. 11: 713–719.
4. **Dovers, S. R.** 1990. Sustainability in context: an Australian perspective. *Environ. Manage*. 14: 297–305.
5. **Kimura, I.** 1988. Aquatic pollution problems in Japan. *Aquat. Toxicol.* 11: 287–301.
6. **Dubos, R.** 1981. *Celebrations of Life*. McGraw-Hill, New York.

Conclusions

It should be mildly disconcerting to those of us who live in coastal areas — and maybe even to all citizens of this technologically advanced country — to find, in a 1987 report of the U.S. Office of Technology Assessment (OTA) titled "Wastes in Marine Environments"[1] that:

> Many municipal and industrial wastes are discharged directly into estuaries and coastal waters [of the U.S.]. More than 1,300 major industrial facilities and 500 municipal sewage treatment plants discharge wastewater effluents directly into estuaries, and an additional 70 municipal plants and about 15 major industrial facilities discharge into coastal waters; only a few pipelines are used to discharge wastewater into the open ocean. Some sewage sludge is discharged through pipelines in southern California and in Boston, although these discharges are scheduled to be terminated.

After that gloomy preamble and a review of status and trends, the OTA assessment reached three general conclusions about the relative states of health of the nation's marine environments:

- Estuaries and coastal waters around the country receive the vast majority of pollutants introduced into marine environments. As a result, many of these waters have exhibited a variety of adverse impacts, and their overall health is declining or threatened.
- In the absence of additional measures, new or continued degradation will occur in many estuaries and some coastal waters around the country during the next few decades (even in some areas that exhibited improvements in the past).
- In contrast, the health of the open ocean generally appears to be better than that of estuaries and coastal waters. Relatively few impacts from waste disposal in the open ocean have been documented, in part because relatively little waste disposal has taken place there and because wastes disposed of there usually are extensively dispersed and diluted.

These conclusions by OTA accord well with those that might be drawn from this book, especially that estuaries and coastal waters are most vulnerable to degradation by human activities.

From the perspective of living marine resources, the single most important polluted-related concern is obviously "the effects of contaminants on fish and shellfish." This is not a simple matter. Information about past and present abundance of resource species is usually incomplete, and the reasons for observed changes in abundance (where such changes have been observed) are often obscure. Cited in the early chapters of this book are *numerous examples of localized effects of pollutants on small segments of fish and shellfish populations, but no specific evidence of widespread damage to major fishery resource populations that can be attributed directly to pollution*. Other factors, such as repeated year-class successes or failures, shifts in geographic distribution of fish populations, or overfishing may cause pronounced changes in fisheries — changes that could obscure any effects of habitat degradation.

Additionally, the magnitude of fluctuations in fish population abundance, due to incompletely known natural environmental causes, makes it extremely difficult to isolate, quantify, and demonstrate the possible role of environmental contaminants in causing such fluctuations.

It may be, of course, that coastal pollution *is* exerting some overall influence on certain resource species, but that this may be masked by increased fishing effort, or by favorable changes in other environmental factors that create a positive effect on abundance that outweighs any negative effects of pollutants. Many experimental studies, particularly those concerned with long-term exposures of fish and shellfish to low levels of contaminants, suggest that some long-term quantitative effects should be felt, but that our statistics, our monitoring, and our population assessments are not adequate to detect them.

It may also be that the effects of present levels of contaminants in coastal environments are not high enough to exert significant or observable impacts on fish populations, except in extremely localized, grossly degraded habitats. Average annual increments to pollution loads may also be small enough to escape detection. Local reductions in abundance could easily be masked in large, often migratory, populations that are being influenced simultaneously by a host of natural environmental factors such as predation, temperature, ocean currents, and food availability.

One additional reason that the effects of pollutants on marine resource species are not easy to demonstrate may be the *resiliency* of estuarine and coastal populations and ecosystems. Human activities (other than overfishing) may not have damaged entire species enough to show significant effects — even though local populations may have been reduced or eliminated, even though eggs and larvae may have been killed, and even though forage organisms may have been affected. Counterbalances acting for the species or the ecosystem, such as the great reproductive potential (fecundity) of most marine species, the great species diversity in most ecosystems, and the adaptive capabilities of many coastal-estuarine species to drastic changes in their physical-chemical environment — all may offset the negative influence of pollutants.

Shifts in species composition of stressed ecosystems have been demonstrated, and some of these changes seem deleterious to economically important fish and shellfish populations. High levels of genetic abnormalities, in the form of chromosomal damage and recurring gross abnormalities, have also been demonstrated in at least a few instances.[2]

Mortalities that seem related to low oxygen content of degraded waters, changes in species composition that seem related to organic composition of bottom sediments, and gross differences in the composition of microbial populations — all may affect ecosystem stability. The combined effects of these negative factors may, however, still be of small enough magnitude as to be hidden among the many variable environmental influences on survival.

The strategic significance of estuaries to the survival and abundance of many fish and shellfish is unquestioned, and it is in estuaries that many of our pollutants are concentrated. It is difficult to understand how estuarine fish and shellfish production systems can avoid collapse, if such systems are being poisoned by pollutants, yet the available evidence does not disclose a general trend toward collapse. Such a generalization, however, should not be used as an argument against vigorous actions toward pollution abatement, even in the absence of incontrovertible evidence of damage to resources. Enough localized and experimental indications of deleterious effects exist to warrant concern.

Action to reduce effects of pollution should precede acquisition of the final proof of damage, but should follow the acquisition of reasonable data (beyond mere extrapolations). Lasting damage to fish populations can occur if action to reduce effects is delayed until absolute proof of damage is acquired, because (to paraphrase an eminent fishery scientist[3]) the final, conclusive proof that a population is being damaged is when it becomes extinct.

But apart from these no-doubt interesting generalizations, what specific conclusions can be distilled from all the verbiage in the preceding chapters? I have identified a few; you may perceive others. Here are mine:

1. Humans place almost impossible conditions for survival on resource populations — preying on adults often at levels beyond which the species can maintain itself (a destructive process called *overfishing*), and simultaneously modifying the chemical environment in some coastal/estuarine sectors enough to reduce growth, reproduction, and survival. Mass mortalities of marine organisms due to extreme natural environmental changes occur, but they may constitute only a small part of the total damage to coastal/estuarine fish populations. Sublethal effects such as spawning failure, poor survival of larvae, reduced growth rates, and increased vulnerability to other environmental limiting factors can have significant effects that may be less apparent. Greater understanding of such sublethal effects is needed to assess fully the influence of any single stressor on living resources. As an aspect of this, the so-called "resiliency" of marine populations is highly variable, since some mass mortalities are followed by very slow recovery of affected populations, whereas in other instances, populations rebound rapidly. The amplitude of population fluctuations can be affected significantly by environmental factors that result in mass mortality. Also, drastic and sometimes long-term changes in community structure may be aftereffects of mass mortalities.

2. Proof of biological damage to resource populations caused by pollutants is very difficult to acquire in the midst of many other factors that influence the abundance of coastal species. Effects may be long term, with average annual increments so small as to escape detection. Experimental information, particularly from long-term chronic exposures to contaminants, suggests that some deleterious effects can be expected and that they are probably occurring, but they escape our methods of detection. Other evidence suggests that natural phenomena are still overriding influences in the survival of coastal populations, but that the dynamic and precarious equilibrium that permits survival may be overturned or distorted by human interference.
3. It is quite likely that threats to public health are developing concomitantly with increasing ocean pollution. Cancer-producing (carcinogenic) products from petroleum and other industrial chemicals occur in the polluted zones and in fish and shellfish living in those zones. There is as yet no clearly demonstrated association between eating seafood products and human cancer occurrences, but the effects may be long term — as may the effects of consumption by humans of low levels of heavy metals and pesticides in the flesh of fish and shellfish.

   We create, then, circular pathways of destruction for marine animals and possibly for ourselves. Pollutants may kill marine species or render them potential hazards to man, instead of available resources. If we persist in using such material as food, it may have low appeal because of odor or flavor, and it may poison us or cause infection. In a sense, humans may be the ultimate bioassay organisms — detecting (by the appearance of abnormalities and disease) over the long term the build-up of carcinogens, mutagens, and other toxic chemicals in the marine environment and marine resource populations. The human species may, in the words of other authors,[4] be "...transformed [by the continued contamination of the environment] into the subjects of a vast experiment in chronic toxicology."
4. The effects of pollutants on marine animals may be expressed in a number of ways, including diminished reproductive activities, damage to genetic material of the egg or embryo with resulting mortality or abnormal development, direct chemical damage to cell membranes or tissues, modification of physiological and biochemical reactions, changes in behavior (often due to chemical damage to sensory equipment), increased infection pressure from facultative microbial pathogens, and reduced resistance to infection. Effects are often expressed as disease — either infectious or noninfectious. Infectious diseases may result from lowered resistance to primary pathogens, from invasion of damaged tissues by facultative (secondary) pathogens, or from proliferation of latent microbial infections. Noninfectious diseases may result from early genetic damage, or from chemical modification of bone and soft tissues, leading to skeletal anomalies and tumors.
5. The capacity of many coastal/estuarine species to accomodate to environmental extremes, including pollutants, has probably been underestimated. Looking specifically at pollutant chemicals, survival may be enhanced by:

   (a) induction of mixed function oxygenases to metabolize hydrocarbons;
   (b) protein binding of heavy metals;
   (c) increased mucus secretion to mitigate chemical stress;
   (d) selection for high immunological competence;

(e) selection of resistant individuals from the population at risk, leading to establishment of survivor populations more tolerant at some life stages to high levels of contaminant chemicals.

Physiological and population responses such as these can lead to prolonged survival and even reproduction in environments that might be considered hostile to aquatic life.

6. Early life history stages of marine animals are usually (but not always) more susceptible to chemical stressors than later stages. The effects may be obvious, resulting in death, or less obvious, such as genetic damage, which may not be expressed until later developmental stages as abnormalities in structure or function. Genetic damage to sex products, embryos, or larvae may be an important and as yet poorly understood consequence of industrial contamination of coastal/estuarine waters.
7. The success of reproduction and survival of year-classes depend on the persistence of a precarious equilibrium state — and man in some instances seems to be disturbing that equilibrium by imposing additional environmental hazards. Evidence exists for localized effects of pollutant stress on fisheries, but as yet there is little specific evidence for widespread damage to major fisheries resource populations resulting from coastal pollution. This may well be because we are unable to separate clearly the effects of pollutant stress from the effects of the many other forms of environmental stressors to which marine populations are subject. Other factors, such as shifts in geographic distribution of fish populations or overfishing, may cause pronounced changes in fisheries — changes that would obscure any effects of habitat degradation in offshore waters.
8. An expanding data base on effects of pollution on living marine resources permits the following assessment:
   - The fact that animals can be injured or killed by exposure to chemical pollutants is clearly supported by thousands of experimental studies.
   - Environmental contaminant levels in certain restricted coastal/estuarine areas are sufficiently high to produce the acute or chronic effects seen in experimental systems.
   - Contaminant levels in tissues of fish have been examined in a number of geographic areas. In many instances, the widespread occurrence of detectable levels of selected contaminants has been demonstrated (PCBs, DDT, mercury, petroleum components).
   - In certain localized areas, tissue levels of contaminants have been recognized in fish that are well above the few legal action levels that exist; toxic effects on the fish at such levels are largely unknown, as are toxic effects of exposure prior to the buildup of measurable tissue levels. Beyond this, the relationship between toxicity and any tissue level has been inadequately explored. High body burdens of contaminants may be sequestered and not affect health, until a level is reached where accommodation is no longer feasible and spillover may occur.
   - Inadequate data are available about the synergistic effects on fish of extremely complex mixtures of contaminant chemicals and dissolved or particulate organics in polluted waters, but enough experimental data exist to state that antagonisms and synergisms in complex mixtures of contaminant chemicals may be important factors in producing any net effect.

Furthermore, even though some experimental data exist about the differential effects of various species, isomers, and substituted forms of contaminant chemicals on fish, the influences of chelation and degradation on biological activity must also be considered.

- Contaminants may have seasonal abundances, with declines during other times of the year. This is because of factors such as spring flushing of river basins or seasonal changes in the biota (e.g., phytoplankton), which tend to concentrate contaminants.
- A much broader perspective focuses on *cumulative effects,* which include not just chemical synergisms, but interactions among fisheries, physical degradation, and chemical contamination. It is possible to consider each item separately, but in the end it is the totality of effects that is significant. As an example, fish do not face just dioxin or mercury, but a whole range of contaminants, and they often must face these materials within highly degraded environments. Most fish are highly mobile, and tend to go from one estuary or coastal area to another. Temporal effects may become apparent when fish visit a highly degraded area, picking up contaminants, but then move on to another area where the effects of the contaminants may be manifested, but in what appears to a "pristine environment."
- Surveys have indicated that some pathological conditions in fish (fin erosion, ulcerations, certain epidermal papillomas, chromosomal damage) can be associated statistically with severe environmental contamination. The statistical relationships have varying degrees of "robustness," and cause-and-effect relationships are difficult to demonstrate for specific contaminants.

9. Given the present rate of human population growth and the likelihood of its increasing impact on coastal/estuarine waters, it is easy to be pessimistic. We may be facing a future in which:

   - Currently isolated zones of severe environmental contamination may gradually expand, intensify, and coalesce, as human population density and industrialization in adjacent land areas increase.
   - Sublethal effects of toxic industrial contaminants may, in presently unknown ways, affect the survival and well-being of resource species, and, indirectly, humans as well.
   - Wild fish and shellfish stocks that survive in the presence of industrial or municipal pollution may be increasingly excluded from markets for public health reasons, since they may carry dangerous levels of toxic chemical contaminants and microbial pathogens.
   - Growing consumer unease and periodic panics about toxic contaminants in seafood may act to reduce the acceptability and hence the market value of products from natural aquatic sources.
   - Fish stocks have collapsed (some spectacularly) in many parts of the world, and no relief seems to be in sight. While fisheries have reached low points in the past, it is, for example, difficult to find a parallel to the present extremely low stock abundance of cod, haddock, yellowtail flounder, and other species in the Northwest Atlantic. If this collapse was so evident, why did it happen? Is it a result of inappropriate interactions between the fisheries managers, scientists, and fishermen? Overfishing and habitat degradation are usually identified as principal culprits; some investigators have

suggested, however, that overfishing is a problem for the short run, but environmental degradation and poor-quality habitats are problems for the long run.

That such a dismal future is close at hand is reinforced by presently emerging environmental news from Europe. Almost daily we learn of the extent to which humans can and will degrade their surroundings, as the horrors of aquatic and terrestrial pollution in former communist bloc Eastern European countries are revealed. We are discovering that the totalitarian regimes of those nations, during the period from 1945 to 1991, tolerated and even encouraged:

–Ocean dumping (including radioactive material) on a scale that trivializes that carried on in North America and Western Europe
–The treatment of great rivers as open sewers, without even the pretense of abatement measures
–The almost total disregard for dangers to public health from effluents and other emissions of industrial complexes

What is unfolding is evidence for almost half a century of deliberate despoliation of the environment, fostered by the overwhelming indifference of government functionaries concerned only with meeting quotas and private gain. The history of that half-century may eventually supply an epic example of the cumulative effects of humans on aquatic resources and their habitats — where rivers have been diverted throughout entire riverine or river-basin systems or have been used primarily for effluent discharge, and where fish have been removed to the point where there are virtually no recruiting stocks left. These are truly cumulative effects, but we in the West have been largely unaware of them.

10. One of the many unfavorable consequences of the human population explosion, accompanied by a counterpart expansion in technology, has been the progressive degradation of coastal/estuarine habitats of fish and shellfish. Humans have only recently been able to make a measurable impact on the marine environment and its inhabitants — by a combination of ruthless exploitation of fish and shellfish stocks and by the simultaneous destruction of their habitats.

Most humans, however, refuse — persistently — to accept the reality that they are part of the planetary ecosystem and, now that they have achieved overwhelming numbers, that their actions can seriously affect all parts of that ecosystem. They have created local and global problems with atmospheric gases; they have destroyed an alarming portion of the earth's forests; they have used major rivers as sewers; and now in this century they have begun to inflict harm on coastal/estuarine resource populations and their habitats. Furthermore, some of the toxic chemical products of their technology, such as DDT and PCBs, have been disseminated globally and can be found in the inhabitants of all oceans.

Rachel Carson, in one of her speeches after the publication of *Silent Spring*[5] said "...all the life of the planet is interrelated; each species has its own ties to others; and all are related to the earth." Most members of the human species have, sadly, rejected and still reject this thesis, even though evidence is accumulating for its validity.

Effects of pullution on living marine resources have received some scrutiny in recent decades, and information is accumulating, but is still insufficient to be very useful in resource management decisions, except as they involve local areas. Evidence exists for localized effects of pollutant stress on fisheries, but as yet there is little specific evidence for widespread damage to major fisheries resource populations resulting from coastal pollution. This may well be because we are unable to separate clearly the effects of pollutant stress from the effects of the many other forms of environmental stressors to which marine populations are subject. Other factors, such as shifts in geographic distribution of fish populations, changes in productive ecosystems, or overfishing, may cause pronounced changes in fisheries — changes that could obscure any effects of localized habitat degradation. It seems, from the evidence presently available, that factors other than pollution seem to be overriding in determining fish abundance, but we lack quantitative data to make positive statements about cause-and-effect relationships of abundance and pollution.

Effective long-term monitoring of stocks and environment must be the basis of any attempt to isolate and identify effects of pollution. A continuous integrated effort in stock assessment, environmental assessment, and experimental studies will be required to understand the role of all environmental stressors — natural and man-induced — in determining the abundance of resource populations. For the present, however, in the absence of a full understanding of the phenomena involved, management decisions affecting coastal/estuarine pollution must be made on the basis of "best available scientific information," just as decisions about allowable resource exploitation are made. In both types of decision processes, a conservative action provides a lower risk of damage and loss than does a more extreme action. Conservatism can be especially significant when decisions are made that might permit pollution to continue or increase, since the long-term effects of existing levels on abundance of resource populations are largely undetermined. In addition to advocating conservatism, we must persist in attempts to quantify the effects of pollution, and to determine the precise pathways through which fishery resources are affected.

REFERENCES

1. Office of Technology Assessment. 1987. *Wastes in Marine Environments*. U.S. Congress, OTA-0–334. U.S. Government Printing Office, Washington, DC.
2. **Longwell, A. C.** 1976. *Chromosome mutagenesis in developing mackerel eggs sampled from the New York Bight*. U.S. Department of Commerce, NOAA Tech. Memo. ERL-MESA-7: 1–61; **Valentine, D. W.** 1975. Skeletal anomalies in marine teleosts, pp. 695–718. in Ribelin, W.E. and G. Migaki (Eds.), *The*

Pathology of Fishes. University of Wisconsin Press, Madison. 1004 pp.; **Valentine, D. W. and K. W. Bridges.** 1969. High incidence of deformities in the serranid fish *Paralabrax nebulifer* from southern California. *Copeia* 1969(3): 637–638.

3. **Gulland, J. A.** 1952. Correlations on fisheries hydrography. Letters to the editor. *J. Cons. Perm. Int. Explor. Mer* 18: 351–353.
4. **Huddle, N. and M. Reich.** 1975. *Island of Dreams: Environmental Crisis in Japan*. Autumn Press, New York, 225 pp.
5. **Carson, R.** 1963. Keynote address. Women's National Book Assoc., Annual Meeting.

Epilogue: The Role of Scientists in Public Environmental Issues

Awareness of environmental degradation that can be attributed to humans has increased significantly in recent decades, especially since the first Earth Day in 1970. The average U.S. citizen has naturally looked to the scientific community for relevant data and for unbiased interpretations of those data. One real problem that has emerged is that there are few unequivocal answers to environmental questions. Most of them must be qualified or surrounded by caveats — a favorite (and correct) response of scientists that is nonetheless very frustrating to everybody else, whether they be regulatory bureaucrats, lawyers, elected officials, newspaper reporters, or private citizens.

Another problem, less frequently encountered but still significant, is the tendency of some scientists to adopt *advocacy* positions, for or against particular points of view or conclusions on environmental issues (but not about the fundamental need for wise environmental management, which they would all advocate). This sometimes subjective behavior can be almost incomprehensible and terribly disconcerting to lay people — scientists arguing among themselves (sometimes publicly) about interpretations of a common data set.

Part of the explanation for this disparity of opinions or interpretations lies in the fact that scientists are human beings, and some find it difficult at times to be absolutely objective about issues — such as certain environmental issues — that are science-based but have high emotional content as well. Depending on their training, their personal philosophies of life, their psychological makeup, pressures from spouses and children, and other stresses of existence, scientists may find themselves, at any given moment, located somewhere along a spectrum of personal positions on any issue ranging from the "neutral observer" to the "extremist," moving through such intermediate positions as "concerned professional" to "advisor" to "advocate" and to "activist." For proper perspective, though, this simple linear progression has to be supplemented by some indication of where most scientists will cluster. In a survey done earlier for other purposes, most scientists classified themselves as "neutral observers" or "concerned professionals" (or as intermediate between the two descriptors), with a precipitous decline in relative numbers of individuals who admitted to being advocates or activists. It seems relevant that many of the scientists who labeled themselves as advocates and beyond tended to be very young or quite elderly; often the older individuals had established credentials in a field of research somewhat tangential to that at issue.

The reason for concern about advocacy here is of course that subjective opinions may obscure or modify the conclusions warranted by the technical information available, or that inadequate or selected data may be extrapolated to provide unwarranted conclusions.

Looking at this spectrum of philosophical positions, it is apparent (at least to me) that the most dangerous excesses are not committed by those activist or extremist scientists who are out on the street with signs — we recognize that they have slipped over the edge and are no longer functioning as professionals. The greatest danger is rather from the previously credible scientist who has become an advocate, maybe by acting as spokesperson for one of an ever-expanding list of organizations with environmental concerns. He or she speaks authoritatively and cleverly in broad and ultimately untenable generalizations; he or she is very selective about the facts that are alluded to in television interviews; and he or she trades heavily on the labels of "scientist" and "expert," which imply membership in an elite class. Resident in this kind of person is the real problem in conveying to the public some feeling for the status of unbiased scientific knowledge about environmental matters.

But there is an important intermediate position here, between the concerned professional and the advocate — that of the *advisor* on environmental matters. This is a position that does not require advocacy, but does require *participation* in public debate, and very careful adherence to a basic ethic of science — truth. It appears almost axiomatic that problems associated with coastal/estuarine pollution deserve examination by the best ecological talent available, and advice to regulatory agencies, politicians, or the public should be based on impartial analysis of all the data available. If these precepts are accepted, then it follows that acceptance of an advisory role by some marine scientists is almost *obligatory*. Responsible scientists should be prepared to say that "...based on the evidence available to me at present, this is what I know and this is what I believe."

Unfortunately, participation in advisory activities can quickly draw even the most cautious and conservative scientists into a legal combat zone where they will be usually ill-equipped by training or experience to compete successfully with the predatory inhabitants of that zone (lawyers, hostile expert witnesses, biased hearing examiners, and opposition agency representatives). Some scientists learn the game rules quickly and find the role of expert witness in environmental matters a challenging one; others are totally repelled by legal maneuvers that may deliberately distort the truth and even bring scientific reputations into question. The reality is, though, that public issues such as ocean pollution are examined from many perspectives — economic, social, and political, as well as scientific — so scientific advice, while extremely valuable, is only one of the bases for decisions. Scientists need to maximize the effects of their findings and conclusions on such decisions, but they must also recognize the complexity of the total process, and the likelihood that their opinions may not always prevail.

With these final observations, the voyage is over. The research data base that is the foundation for the text will be, I am sure, exploited from other perspectives by other writers, who will also have the advantage of access to new information that is being published in overwhelming quantity. I have reached conclusions and drawn inferences that seem to be consistent with the available scientific literature — conclusions and inferences that may be supported or refuted in someone else's book a decade from now. Producing this volume has been an exhilarating long-term experience, extending as it has over a span of almost 20 years (with occasional sterile periods). My hope is that the document will provide readers with a small window on some of the principal resource problems that have been created by ocean pollution.

INDEX

Index

A

B

C

D

E

F

G

H

I

J

K

L

N

O

P

R

S

T